Ratios and Proportions

A complete workbook with lessons and problems

By Maria Miller

Contents

Preface 5

Introduction 7

Helpful Resources on the Internet 8

Ratios and Rates 13

Solving Problems Using Equivalent Rates 16

Solving Proportions: Cross Multiplying 18

Why Cross Multiplying Works 24

Unit Rates 25

Proportional Relationships 30

Graphing Proportional Relationships–More Practice 36

More on Proportions 38

Scaling Figures 42

Floor Plans 48

Maps 52

Significant Digits 58

Review 60

Answers 65

Appendix: Common Core Alignment 89

Preface

Hello! I am Maria Miller, the author of this math book. I love math, and I also love teaching. I hope that I can help you to love math also!

I was born in Finland, where I also grew up and received all of my education, including a Master's degree in mathematics. After I left Finland, I started tutoring some home-schooled children in mathematics. That was what sparked me to start writing math books in 2002, and I have kept on going ever since.

In my spare time, I enjoy swimming, bicycling, playing the piano, reading, and helping out with Inspire4.com website. You can learn more about me and about my other books at the website MathMammoth.com.

This book, along with all of my books, focuses on the conceptual side of math... also called the "why" of math. It is a part of a series of workbooks that covers all math concepts and topics for grades 1-7. Each book contains both instruction and exercises, so is actually better termed *worktext* (a textbook and workbook combined).

My lower level books (approximately grades 1-5) explain a lot of mental math strategies, which help build number sense — proven in studies to predict a student's further success in algebra.

All of the books employ visual models and exercises based on visual models, which, again, help you comprehend the "why" of math. The "how" of math, or procedures and algorithms, are not forgotten either. In these books, you will find plenty of varying exercises which will help you look at the ideas of math from several different angles.

I hope you will enjoy learning math with me!

Introduction

Ratios and Proportions Workbook covers the concept of the ratio of two quantities. From this concept, we develop the related concepts of a rate (so much of one thing per so much of another thing) and a proportion (an equation of ratios). We also study how tables of equivalent ratios can help to solve problems with rates, and how cross-multiplying can help to solve problems with proportions.

The lesson *Unit Rates* defines the concept of the unit rate, shows how to calculate one, and gives practice at doing so, including practice with complex fractions. We also consider rates as two quantities that vary, graph the corresponding equation in the coordinate grid, and tie in the concept of unit rate with the concept of slope.

The concept of direct variation is introduced in the lesson *Proportional Relationships*. Writing and graphing equations gives a visual understanding of proportionality. In two following lessons on proportions, students also practice solving rate problems in different ways, using the various methods they have learned throughout the workbook.

The lessons *Scaling Figures, Floor Plans, and Maps* give useful applications and more practice to master the concepts of proportions.

Before the *Review* there is also an optional lesson, *Significant Digits*, that deals with the concept of the accuracy of a measurement and how it limits the accuracy of the solution. It is optional because significant digits is not a standard topic for seventh grade, yet the concept in it is quite important, especially in science.

I wish you success in teaching math!

Maria Miller, the author

Helpful Resources on the Internet

Use these free online games and resources to supplement the "bookwork" as you see fit.

Ratios and Proportions—Video Lessons by Maria
A set of free videos about ratios, rates, and proportions—by the author.
http://www.mathmammoth.com/videos/proportions.php

RATIOS AND RATES

Ratio Pairs Matching Game
Match cards representing equivalent ratios.
Easy: http://nrich.maths.org/4824 Challenge: http://nrich.maths.org/4821

All About Ratios - Quiz
An interactive five-question quiz about equivalent ratios presented with pictures
http://math.rice.edu/~lanius/proportions/quiz1.html

Rate Module from BrainingCamp
A comprehensive interactive lesson on the concepts of ratio, rate, and constant speed (for 6th and 7th grades). Includes an animated lesson, a virtual manipulative, and questions and problems to solve.
http://www.brainingcamp.com/lessons/rates/

Ratios Activity from BBC Bitesize
An animated tutorial about dividing in a given ratio and scale models with some quiz questions along the way.
http://www.bbc.co.uk/education/guides/znnycdm/activity

Self-Check Quiz from Glencoe
A five-question multiple-choice quiz about comparing with ratios and rates. By reloading the page you will get different questions.
http://www.glencoe.com/sec/math/studytools/cgi-bin/msgQuiz.php4?isbn=1-57039-855-0&chapter=8&lesson=1

Ratio Quiz from BBC Skillswise
A multiple-choice quiz about the concept of ratio. You can take the quiz online or download it as a PDF or doc file.
http://www.bbc.co.uk/skillswise/quiz/ma19rati-e1and2-quiz

Ratio Quiz from Syvum
A 10-question online quiz about ratios and problem solving.
http://www.syvum.com/cgi/online/mult.cgi/gmat/math_review/arithmetic_5.tdf?0

Unit Rates Involving Fractions
Practice computing rates associated with ratios of fractions or decimals in this interactive activity.
https://www.khanacademy.org/math/cc-seventh-grade-math/cc-7th-ratio-proportion/cc-7th-rates/e/rate_problems_1

Ratio Word Problems
Reinforce your ratios skills with these interactive word problems.
https://www.khanacademy.org/math/pre-algebra/pre-algebra-ratios-rates/pre-algebra-ratio-word-problems/e/ratio_word_problems

Comparing Rates
Practice completing rate charts in this interactive online activity.
https://www.khanacademy.org/math/pre-algebra/pre-algebra-ratios-rates/pre-algebra-rates/e/comparing-rates

Free Ride
An interactive activity about bicycle gear ratios. Choose the front and back gears, which determines the gear ratio. Then choose a route, pedal forward, and make sure you land exactly on the five flags.
http://illuminations.nctm.org/ActivityDetail.aspx?ID=178

Exploring Rate, Ratio and Proportion (Video Interactive)
The video portion of this resource illustrates how these math concepts play a role in photography. The interactive component allows students to explore ratio equivalencies by enlarging and reducing images.
http://www.learnalberta.ca/content/mejhm/index.html?l=0&ID1=AB.MATH.JR.NUMB&ID2=AB.MATH.JR.NUMB.RATE

Three-Term Ratios
Practice the equivalency of ratios by filling in the missing numbers in three-term ratios
(for example, 2:7:5 = __ : 105 : ___) where the numbers represent the amounts of three colors in different photographs. Afterwards you get to assemble a puzzle from the nine photographs.
http://www.learnalberta.ca/content/mejhm/index.html?
l=0&ID1=AB.MATH.JR.NUMB&ID2=AB.MATH.JR.NUMB.RATE&lesson=html/object_interactives/3_term_ratio/use_it.html

If the two links above don't work, use this link:
http://www.learnalberta.ca/content/mejhm/index.html?l=0&ID1=AB.MATH.JR.NUMB
First choose Rate/Ratio/Proportion, and then either *Exploring Rate, Ratio, and Proportion* or *3-Term Ratios*.

PROPORTIONS

Ratios and Proportions
A tutorial with interactive practice exercises about ratios and proportions.
https://www.wisc-online.com/learn/formal-science/mathematics/gem2004/ratios-and-proportions

Solving Proportions Practice
In this interactive practice, you can choose to show a hint "vertically", "horizontally", or algebraically.
http://www.xpmath.com/forums/arcade.php?do=play&gameid=97

Solving Proportions
Practice solving basic proportions with this interactive exercise from Khan Academy.
https://www.khanacademy.org/exercise/proportions_1

Proportions Quiz
Use this multiple-choice self-check quiz to test your knowledge about proportions.
http://www.glencoe.com/sec/math/studytools/cgi-bin/msgQuiz.php4?isbn=0-07-829633-1&chapter=7&lesson=3&headerFile=4

Proportions: Short Quiz
This short multiple-choice quiz reinforces basic proportions skills.
http://www.phschool.com/webcodes10/index.cfm?wcprefix=bja&wcsuffix=0602&area=view

Challenge Board
Choose questions from the challenge board about rates, ratios, and proportions.
http://www.quia.com/cb/158527.html
http://www.quia.com/cb/101022.html

Write a Proportion to a Problem
Practice writing proportions to describe real-world situations in this interactive exercise.
https://www.khanacademy.org/math/pre-algebra/pre-algebra-ratios-rates/pre-algebra-write-and-solve-proportions/e/writing_proportions

Proportion Word Problems
Practice setting up and solving proportions to solve word problems in this interactive online activity.
https://www.khanacademy.org/math/pre-algebra/pre-algebra-ratios-rates/pre-algebra-write-and-solve-proportions/e/constructing-proportions-to-solve-application-problems

Rags to Riches—Proportions
Solve proportions and advance towards more and more difficult questions.
http://www.quia.com/rr/35025.html

How Much Is a Million?
This is a lesson plan for a hands-on activity where students count grains of rice in a cup, weigh that amount of rice, and then build a proportion to figure out the weight of 1 million grains of rice.
http://illuminations.nctm.org/Lesson.aspx?id=2674

PROPORTIONAL RELATIONSHIPS

Proportional Relationships and Graphs
Practice plotting points on a graph and reading graphs in this interactive online activity.
http://www.buzzmath.com/Docs#CC07E11805

Proportional Relationships Activity
Answer true or false questions, practice reading charts, and more in this interactive online activity.
http://www.buzzmath.com/Docs#CC07E11806

Proportional Relationships
Practice telling whether or not the relationship between two quantities is proportional by reasoning about equivalent ratios.
https://www.khanacademy.org/math/pre-algebra/pre-algebra-ratios-rates/pre-algebra-proportional-rel/e/analyzing-and-identifying-proportional-relationships

Proportional Relationships: Graphs
Practice telling whether or not the relationship between two quantities is proportional by looking at a graph of the relationship.
https://www.khanacademy.org/math/cc-seventh-grade-math/cc-7th-ratio-proportion/cc-7th-graphs-proportions/e/analyzing-and-identifying-proportional-relationships-2

Interpreting Graphs of Proportional Relationships
Practice reading and analyzing graphs of proportional relationships in this interactive online exercise.
https://www.khanacademy.org/math/cc-seventh-grade-math/cc-7th-ratio-proportion/cc-7th-equations-of-proportional-relationships/e/interpreting-graphs-of-proportional-relationships

Writing Proportional Equations
Practice writing equations to describe proportional relationships in this interactive online activity.
https://www.khanacademy.org/math/cc-seventh-grade-math/cc-7th-ratio-proportion/cc-7th-equations-of-proportional-relationships/e/writing-proportional-equations

SCALE DRAWINGS AND MAPS

Similar Figures Activity
Practice finding the missing length, answer questions about proportions, and more with this interactive activity.
http://www.buzzmath.com/Docs#F6KME681

Similar Shapes Exercises
Answer questions about the scale factors of lengths, areas, and volumes of similar shapes.
http://www.transum.org/Maths/Activity/Similar/

Scale Drawings Exercise
Measure line segments and angles in geometric figures, including interpreting scale drawings.
http://www.transum.org/software/Online_Exercise/ScaleDrawing/

Scale Drawings Quizzes
Interactive self-check quizzes about scale drawings. By reloading the page you will get different questions.
http://www.glencoe.com/sec/math/studytools/cgi-bin/msgQuiz.php4?isbn=0-07-860467-2&chapter=4&lesson=1&headerFile=4

http://www.glencoe.com/sec/math/studytools/cgi-bin/msgQuiz.php4?isbn=0-02-833050-1&chapter=8&lesson=3

Ratio and Scale
An online unit about scale models, scale factors, and maps with interactive exercises and animations.
http://www.absorblearning.com/mathematics/demo/units/KCA024.html

Use Proportions to Solve Problems Involving Scale Drawings
A set of word problems. You can choose how they are presented: as flashcards, as a quiz where you match questions and answers, as a multiple choice quiz, or a true/false quiz. You can also play a game (Jewels) .
http://www.cram.com/flashcards/use-proportions-to-solve-problems-involving-scale-drawings-3453121

Scale Drawings - Problem Solving and Constructing Scale Drawings Using Various Scales
A comprehensive lesson with several worked out examples concerning scale drawings.
http://www.ck12.org/user:c2ZveDJAb3N3ZWdvLm9yZw../book/Oswego-City-School-District---Grade-7-Common-Core/section/12.0/

Constructing Scale Drawings
Practice making scale drawings on an interactive grid. The system includes hints and the ability to check answers.
https://www.khanacademy.org/math/cc-seventh-grade-math/cc-7th-geometry/cc-7th-scale-drawings/e/constructing-scale-drawings

Interpreting Scale Drawings
Solve word problems involving scale drawings in an online practice environment.
https://www.khanacademy.org/math/cc-seventh-grade-math/cc-7th-geometry/cc-7th-scale-drawings/e/interpreting-scale-drawings

Scale Drawings and Maps Quiz
Answer questions about scales on maps and scale drawings in these five self-check word problems.
http://www.math6.org/ratios/8.6_quiz.htm

Maps
A tutorial with worked out examples and interactive exercises about how to calculate distances on the map or in real life based on the map's scale.
http://www.cimt.org.uk/projects/mepres/book7/bk7i19/bk7_19i3.htm

Short Quiz on Maps
Practice map-related concepts in this multiple-choice quiz.
http://www.proprofs.com/quiz-school/story.php?title=map-scales

SIGNIFICANT DIGITS

Sig Fig Rules
Drag Sig J. Fig to cover each significant digit in the given number.
http://www.sigfig.dreamhosters.com/

Practice on Significant Figures
A multiple-choice quiz that also reminds you of the rules for significant digits.
http://www.chemistrywithmsdana.org/wp-content/uploads/2012/07/SigFig.html

Significant digits quiz
A 10-question multiple-choice quiz about significant digits.
http://www.quia.com/quiz/114241.html?AP_rand=1260486279

Ratios and Rates

A **ratio** is a comparison of two numbers, or quantities, using division.

For example, to compare the hearts to the stars in the picture, we say that the ratio of hearts to stars is 5:10 (read "five to ten").

The two numbers in the ratio are called the **first term** and the **second term** of the ratio. The order in which these terms are mentioned does matter! For example, the ratio of stars to hearts is *not* the same as the ratio of hearts to stars. The former is 10:5 and the latter is 5:10.

We can write this ratio in several different ways:

- The ratio of hearts to stars is 5:10.
- The ratio of hearts to stars is $\frac{5}{10}$.
- The ratio of hearts to stars is 5 to 10.
- For every five hearts, there are ten stars.

Note that we are not comparing two numbers to determine which one is greater (as in 5 < 10). The comparison is relative as in a multiplication problem. For example, the ratio 5:10 can be simplified to 1:2, and it indicates to us that there are twice as many stars as there are hearts.

We **simplify ratios** in exactly the same way we simplify fractions.

Example 1. In the picture at the right, the ratio of hearts to stars is 12:16. We can simplify that ratio to 6:8 and even further to 3:4. These three ratios (12:16, 6:8, and 3:4) are called **equivalent ratios**.

The ratio that is simplified to lowest terms, 3:4, tells us that for every three hearts, there are four stars.

1. Write the ratio and then simplify it to lowest terms.

The ratio of triangles to diamonds is _____ : _____ = _____ : _____ .

In this picture, there are _____ triangles to every _____ diamonds.

2. **a.** Draw a picture with pentagons and circles so that the ratio of pentagons to the total of all the shapes is 7:9.

b. What is the ratio of circles to pentagons?

3. **a.** Draw a picture in which (1) there are three diamonds for every five triangles, and (2) there is a total of 9 diamonds.

b. Write the ratio of all the diamonds to all the triangles, and simplify this ratio to lowest terms.

4. Write the equivalent ratios.

a. 5 to 45 = 1 to _____	**b.** 3 : _____ = 9 : 60	**c.** 280 : 420 = 2 : _____	**d.** $\frac{5}{13} = \frac{\boxed{}}{65}$

We can also form **ratios using quantities that have units**. If the units are the same, they cancel.

Example 2. Simplify the ratio 250 g : 1.5 kg.

First we convert 1 kg to grams and then simplify: $\dfrac{250 \text{ g}}{1.5 \text{ kg}} = \dfrac{250 \text{ g}}{1{,}500 \text{ g}} = \dfrac{250}{1{,}500} = \dfrac{1}{6}$.

5. Use a fraction line to write ratios of the given quantities as in the example. Then simplify the ratios.

a. 5 kg and 800 g $\dfrac{5 \text{ kg}}{800 \text{ g}} =$	**b.** 600 cm and 2.4 m
c. 1 gallon and 3 quarts	**d.** 3 ft 4 in and 1 ft 4 in

We can generally **convert ratios with decimals or fractions into ratios of whole numbers**.

Example 3. Because we can multiply both terms of the ratio by 10, $\dfrac{1.5 \text{ km}}{2 \text{ km}} = \dfrac{15 \text{ km}}{20 \text{ km}}$.

Then: $\dfrac{15 \text{ km}}{20 \text{ km}} = \dfrac{15}{20} = \dfrac{3}{4}$. So the ratio 1.5 km : 2 km is equal to 3:4.

You can also see that the ratio is 3:4 by noticing that both 1.5 km and 2 km are evenly divisible by 500 m.

Example 4. Simplify the ratio ¼ mile to 5 miles.

First, ¼ mi : 5 mi = ¼ : 5. Multiplying both terms of the ratio by 4, we get ¼ : 5 = 1:20.

6. Use a fraction line to write ratios of the given quantities. Then simplify the ratios to integers.

a. 5.6 km and 3.2 km	**b.** 0.02 m and 0.5 m
c. 1.25 m and 0.5 m	**d.** 1/2 L and 7 1/2 L
e. 1/2 mi and 3 1/2 mi	**f.** 2/3 km and 1 km

If the two terms in a ratio have *different* units, then the ratio is also called a **rate.**

Example 5. The ratio "5 miles to 40 minutes" is a rate that compares the quantities "5 miles" and "40 minutes," perhaps for the purpose of giving us the speed at which a person is running.

We can write this rate as 5 miles : 40 minutes or $\dfrac{5 \text{ miles}}{40 \text{ minutes}}$ or 5 miles *per* 40 minutes.

The word "per" in a rate signifies the same thing as a colon or a fraction line.

This rate can be simplified: $\dfrac{5 \text{ miles}}{40 \text{ minutes}} = \dfrac{1 \text{ mile}}{8 \text{ minutes}}$. The person runs 1 mile in 8 minutes.

Example 6. Simplify the rate "15 pencils per 100¢." Solution: $\dfrac{15 \text{ pencils}}{100¢} = \dfrac{3 \text{ pencils}}{20¢}$.

7. Write each rate using a colon, the word "per," or a fraction line. Then simplify it.

    a. Jeff swims at a constant speed of 1,200 ft in 15 minutes.

    b. A car can travel 54 miles on 3 gallons of gasoline.

8. Fill in the missing numbers to form equivalent rates.

a. $\dfrac{1/2 \text{ cm}}{30 \text{ min}} = \dfrac{}{1 \text{ h}} = \dfrac{}{15 \text{ min}}$ **b.** $\dfrac{\$88.40}{8 \text{ hr}} = \dfrac{}{2 \text{ hr}} = \dfrac{}{10 \text{ hr}}$

9. Simplify these rates. Don't forget to write the units.

a. 280 km per 7 hours	**b.** 2.5 inches : 1.5 minutes

10. A car is traveling at a constant speed of 72 km/hour. Fill in the table of equivalent rates: each pair of numbers in the table (distance/time) forms a rate that is equivalent to the rate 72 km/hour.

Distance (km)							
Time (min)	10	30	40	50	60	90	100

11. Eight pairs of socks cost $20. Fill in the table of equivalent rates.

Cost ($)							
Pairs of socks	1	2	4	6	7	8	9	10

Solving Problems Using Equivalent Rates

Example 1. It took Jack 1 1/2 hours to paint 24 feet of fence. Painting at the same speed, how long will it take him to paint the rest of the fence, which is 100 feet long?

In this problem, we see a rate of 24 ft per 1 1/2 hours. There is another rate, too: 100 ft per an unknown amount of time. These two are equivalent rates. We can use a table of equivalent rates to solve the problem.

Amount of fence (ft)	24	4	48	96	100
Time (minutes)	90	15	180	360	375

(1) We figure that Jack can paint 4 ft of fence in 15 minutes (by dividing the terms in the original rate by 6).

(2) Next we double both terms in the original rate of 24 ft/90 min to get the rate 48 ft/180 min.

(3) Then we double that rate to get 96 ft/360 min.

(4) Lastly, since 100 ft = 96 ft + 4 ft, we add the corresponding times to get
360 min + 15 min = 375 min, or 6 hours 15 minutes.

Example 2. You get 20 erasers for $1.90. How much would 22 erasers cost?

We can solve the problem using a table of equivalent ratios:

Price	$1.90	$0.19	$2.09
Erasers	20	2	22

Twenty erasers cost $1.90, so 2 erasers would cost 1/10 that much, or $0.19. Lastly, we find the cost of 22 erasers, which is the cost of 20 + 2 erasers, or $1.90 + $0.19 = $2.09.

1. Fill in the tables of equivalent rates.

a. Distance	15 km			
Time	3 hr	1 hr	15 min	45 min

b. Pay	$6			
Time	45 min	15 min	1 hr	1 hr 45 min

2. Fill in the missing terms in these equivalent rates.

a. $\dfrac{3 \text{ in}}{8 \text{ ft}} = \dfrac{}{2 \text{ ft}} = \dfrac{}{12 \text{ ft}} = \dfrac{}{20 \text{ ft}}$

b. $\dfrac{115 \text{ words}}{2 \text{ min}} = \dfrac{}{1 \text{ min}} = \dfrac{}{3 \text{ min}}$

3. Jake can ride his bicycle 8 miles in 14 minutes. At the same constant speed, how long will he take to go 36 miles? Fill in the equivalent rates below.

$\dfrac{8 \text{ miles}}{14 \text{ minutes}} = \dfrac{4 \text{ miles}}{\boxed{} \text{ minutes}} = \dfrac{36 \text{ miles}}{\boxed{} \text{ minutes}}$

4. Larry earns $90 for seven hours of work. In how many hours will he earn $600?

Earnings							
Work Hours							

Example 3. Jake can ride his bike 20 miles in 45 minutes. Riding at the same constant speed, how far could he ride in 1 hour?

Remember that we can <u>multiply or divide both terms of a rate</u> by the same number to form another, equivalent rate. You have used this same idea in the past with equivalent fractions.

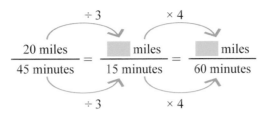

It is not easy to go directly from the rate of 20 miles in 45 minutes to the equivalent rate for 60 minutes. However, we can easily find the rate for 15 minutes: simply divide both 20 miles and 45 minutes by 3. In case you are stumped by 20 ÷ 3, remember that it is easy to solve when you think of it as a fraction: 20/3 = 6 2/3. We get the rate 6 2/3 miles per 15 minutes.

Then, we multiply both terms of that rate by 4. Again, don't be intimidated by the fraction: 4 · (6 2/3) = 4 · 6 + 4 · (2/3) = 24 + 8/3 = 26 2/3. So Jake can ride 26 2/3 miles in 1 hour.

5. A car can travel 45 miles on 2 gallons of gasoline. How many gallons of gasoline would the car need for a trip of 60 miles?

6. **a.** A train travels at a constant speed of 80 miles per hour. How far will it go in 140 minutes?

$$\frac{80 \text{ miles}}{60 \text{ min}} =$$

b. Is this equal to the rate of traveling 50 miles in 40 minutes?

7. You get 30 pencils for $4.50. Is that equal to the rate of 50 pencils for $7.25? You can make a table of equivalent rates to help yourself.

Price							
Pencils							

8. In a poll that interviewed 1,000 people about their favorite color, 640 people said they liked blue.

a. Simplify this ratio to lowest terms.

b. Assuming the same ratio holds true in another group of 125 people, how many of those people can we expect to like blue?

Solving Proportions: Cross Multiplying

A **proportion** is an equation where one ratio is set equal to another ratio.

The equations below are examples of proportions. Notice that, in each one, one ratio is equal to another.

$$\frac{20 \text{ km}}{30 \text{ min}} = \frac{55 \text{ km}}{x} \qquad\qquad \frac{7}{8} = \frac{21}{24} \qquad\qquad \frac{12.2}{T} = \frac{15.6}{8.7}$$

You can solve proportions in several different ways. For example, you can think of them as equivalent fractions or use a table of equivalent ratios like we did in the previous lesson. However, the most common way of solving proportions is **cross-multiplying**. It is the most efficient method to use when the numbers are not simple.

Example 1. Solve the proportion below.

$$\frac{7}{16} = \frac{143}{x}$$
This is the proportion.

$$\frac{7}{16} \diagdown \frac{143}{x}$$
First, **cross-multiply** (multiply crisscross) as indicated by the lines in the picture: $7 \cdot x$ and $16 \cdot 143$.
(Realize that what you are really doing is multiplying both sides by 16 and also by x, but the 16 cancels on the left and the x cancels on the right.)

$$7x = 16 \cdot 143$$
After cross-multiplying we get this equation.

$$7x = 2{,}288$$
Here, we have multiplied $16 \cdot 143 = 2{,}288$ on the right side. Next we will divide both sides of the equation by 7.

$$\frac{7x}{7} = \frac{2{,}288}{7}$$
In this last step, the left side simplifies to x. On the right side, we simply perform the division.

$$x \approx 326.86$$
This is the final answer.

You can use a calculator for all the problems in this lesson.

1. Solve the proportions step-by-step. Round your answers to the nearest tenth.

a. $\dfrac{15}{32} = \dfrac{67}{M}$	First cross-multiply.	**b.** $\dfrac{7}{146} = \dfrac{38}{S}$	First cross-multiply.
$15M =$	This is the equation you get after cross-multiplying.	$=$	This is the equation you get after cross-multiplying.
$=$	In this step, calculate what is on the right side.	$=$	Calculate $146 \cdot 38$.
$\underline{\qquad\qquad} = \underline{\qquad\qquad}$	In this step, divide both sides of the equation by ____.	$\underline{\qquad\qquad} = \underline{\qquad\qquad}$	In this step, divide both sides of the equation by ____.
$M =$	This is the final answer.	$S =$	This is the final answer.

2. Solve the proportions step-by-step. Round your answers to the nearest tenth.

a. $\dfrac{1.2}{4.5} = \dfrac{G}{7.0}$ First cross-multiply.

(You may want to flip the sides so the variable is on the left.)

Calculate what is on the right side.

$\underline{\hphantom{xxxxx}} = \underline{\hphantom{xxxxx}}$ In this step, divide both sides of the equation by _____.

$G \ =$ This is the final answer.

b. $\dfrac{4.3}{C} = \dfrac{10.0}{17.0}$ First cross-multiply.

(You may want to flip the sides so the variable is on the left.)

Calculate what is on the right side.

$\underline{\hphantom{xxxxx}} = \underline{\hphantom{xxxxx}}$ In this step, divide both sides of the equation by _____.

$R \ =$ This is the final answer.

3. Solve the following proportions by using cross-multiplication. Give your answers to the nearest hundredth.

a. $\dfrac{T}{25} = \dfrac{15}{3}$

b. $\dfrac{17}{214} = \dfrac{2}{M}$

Example 2. An International cell phone call costs $41 per hour. How much would a call of 27 minutes cost?

The problem involves two rates: $41 per 1 hour and an unknown cost per 27 minutes. To write a proportion, use a variable for the unknown and set those two rates to be equal. Note also that we need to change the 1 hour to 60 minutes so that the two amounts of time use the same unit.

$$\frac{\$41}{60 \text{ min}} \diagdown \frac{M}{27 \text{ min}}$$

This is the proportion. We cross-multiply as indicated by the lines: $60 \cdot M$ and $41 \cdot 27$.

$$\$41 \cdot 27 \text{ min} = 60 \text{ min} \cdot M$$

This is the equation we get after cross-multiplying. We will keep the units "min" and "$" in the equation.

$$60 \text{ min} \cdot M = \$41 \cdot 27 \text{ min}$$

We switched the sides to put the variable on the left.

$$60 \text{ min} \cdot M = \$1107 \cdot \text{min}$$

Here, we have multiplied $27 \cdot 41 = 1107$ on the right side. Notice, the unit "min" is still kept in the equation.

$$\frac{\cancel{60 \text{ min}} \cdot M}{\cancel{60 \text{ min}}} = \frac{\$1107 \cdot \cancel{\text{min}}}{60 \cancel{\text{min}}}$$

Now we divide both sides of the equation by the quantity "60 min" — not only by 60, but by "60 min." On the left side, "60 min" and "60 min" cancel out. On the right side, the units "min" cancel. Notice that the unit "$" does not cancel out, which means our final answer will be in dollars.

$$M = \$18.45$$

The 27-minute international cell phone call would cost $18.45.

Here is another example problem that shows you how to work with the units while solving the equation.

Example 3. A jet airplane flies 5,600 km in 4 hours 40 minutes. Maintaining that same speed, how long will it take to fly 2,150 km?

Again, the problem involves two equal rates, so we can solve it by using a proportion. First we choose the variable T for the unknown time. Next we write an equals sign between the two rates, 5,600 km per 280 minutes and 2,150 km per T. Then we also convert the flying time of 4 hours and 40 minutes into 280 minutes.

$$\frac{5,600 \text{ km}}{280 \text{ min}} = \frac{2,150 \text{ km}}{T}$$

Here is the proportion. Notice that the distances are on top and the times are on the bottom — they "match."

$$\frac{5,600 \text{ km}}{280 \text{ min}} \diagdown \frac{2,150 \text{ km}}{T}$$

Now we cross-multiply: $5,600 \text{ km} \cdot T$ and $2,150 \text{ km} \cdot 280 \text{ min}$. Notice that each of these multiplications involves two *different* quantities: a <u>distance</u> by a <u>time</u>. If we get a <u>distance</u> times a <u>distance</u> or a <u>time</u> times a <u>time</u>, then the proportion is set up incorrectly.

$$5,600 \text{ km} \cdot T = 2,150 \text{ km} \cdot 280 \text{ min}$$

This is what we have after cross-multiplying. We keep the units "min" and "km" in the equation.

$$5,600 \text{ km} \cdot T = 602,000 \text{ km} \cdot \text{min}$$

Here we have multiplied $2,150 \cdot 280 = 602,000$ on the right side. The units "km" and "min" have been placed *behind* that number, and they are still multiplied (km · min)!

$$\frac{\cancel{5,600 \text{ km}} \cdot T}{\cancel{5,600 \text{ km}}} = \frac{602,000 \cancel{\text{ km}} \cdot \text{min}}{5,600 \cancel{\text{ km}}}$$

Here we divide both sides of the equation by 5,600 km. On the left side, "5,600 km" and "5,600 km" cancel out. On the right side, the units "km" cancel. The unit "min" does not cancel out.

$$T = 107.5 \text{ min}$$

This is the final answer. The airplane takes 107.5 minutes, or 1 hour 47.5 minutes, to fly 2,150 km.

4. For each problem below, write a proportion and solve it. Carry the units through your calculation. Don't forget to check that your answer is reasonable.

a. To paint 700 m^2 May needs 85 liters of paint. How much paint does she need to paint 240 m^2?	**b.** A car can travel 56.0 miles on 3.2 gallons of gasoline. How far can it go with 7.9 gallons of gasoline?

5. Jack and Noah each wrote a different proportion for the problem below. Solve both proportions. Carry the units through your calculations. Which proportion gives the correct answer—or are both correct?

A car travels 154 miles on 5.5 gallons of gasoline. How many gallons would it need to travel 900 miles?	
Jack's proportion: $$\frac{154 \text{ mi}}{5.5 \text{ gal}} = \frac{900 \text{ mi}}{x}$$	Noah's proportion: $$\frac{5.5 \text{ gal}}{154 \text{ mi}} = \frac{x}{900 \text{ mi}}$$

In the previous problem, you saw that two proportions that were set up differently could both yield the correct answer. How can that be? The answer lies in the equations you get after cross-multiplying: they were identical.

For word problems that call for a proportion, there are several ways to set up the proportion correctly and several ways to set it up incorrectly. You can spot the incorrect ways by two facts:

(1) You end up multiplying quantities with the same unit, such as the amount of fuel by the amount of fuel, or the price by the price.

(2) Your answer is not reasonable.

Example 4. If 7 lb of chicken costs \$12, how much would 100 lb cost?

Each of the proportions below will give you a correct answer because each one of them leads to the equation with 7 lb $\cdot$ x on one side and \$12 $\cdot$ 100 lb on the other side.

$$\frac{7 \text{ lb}}{\$12} = \frac{100 \text{ lb}}{x} \qquad \frac{\$12}{7 \text{ lb}} = \frac{x}{100 \text{ lb}} \qquad \frac{\$12}{x} = \frac{7 \text{ lb}}{100 \text{ lb}} \qquad \frac{x}{\$12} = \frac{100 \text{ lb}}{7 \text{ lb}}$$

The proportions below will *not* work. In some of them, you end up multiplying pounds by pounds. In others, you end up multiplying x by 100 lb and getting the answer $x = \$0.84$, which is not reasonable. When things like that happen, it means the that two ratios got mixed up. Always check that your answer is reasonable.

$$\frac{7 \text{ lb}}{\$12} = \frac{x}{100 \text{ lb}} \qquad \frac{100 \text{ lb}}{\$12} = \frac{x}{7 \text{ lb}} \qquad \frac{\$12}{x} = \frac{100 \text{ lb}}{7 \text{ lb}} \qquad \frac{7 \text{ lb}}{x} = \frac{100 \text{ lb}}{\$12}$$

6. Choose the proportion that is set up correctly and solve it ("s" means seconds).

a. $\qquad \dfrac{3 \text{ s}}{23 \text{ m}} = \dfrac{x}{7 \text{ s}}$

b. $\qquad \dfrac{23 \text{ m}}{3 \text{ s}} = \dfrac{x}{7 \text{ s}}$

7. Jane wrote a proportion to solve the following problem. Explain what is wrong with her proportion and correct it. Then solve the corrected proportion. Also, <u>check that your answer is reasonable</u>.

Twenty kilograms of premium dog food cost $51. How much would 17 kg cost?	Jane's proportion: $$\frac{20 \text{ kg}}{\$51} = \frac{x}{17 \text{ kg}}$$

8. For each problem below, write a proportion and solve it. Remember to check that your answer is reasonable.

a. A 5.0 lb bag of fertilizer covers 1,000 square feet of lawn. How much fertilizer would you need for a rectangular 30 ft by 24 ft lawn?	**b.** To paint 700 m^2 you need 85 liters of paint. How much area would 6 liters of paint cover?

Why Cross-Multiplying Works

Recall that if we multiply both sides of an equation by the same number, the two sides are still equal.

In a proportion, we have two different numbers in the denominators. We can first multiply both sides of the proportion by the one denominator and then by the other, or we can cross-multiply. Cross-multiplying is in reality just a *shortcut* for doing those two separate multiplications at the same time.

Let's solve the proportion below by multiplying both sides first by the one denominator, then by the other.

$$\frac{6}{13} = \frac{93}{T}$$

First we multiply both sides by 13.

$$\frac{6}{\cancel{13}} \cdot \cancel{13} = \frac{93}{T} \cdot 13$$

Now, on the left side, the 13 in the denominator cancels the other 13.

$$6 = \frac{93}{T} \cdot 13$$

The next step is to multiply both sides by T.

$$6T = \frac{93}{\cancel{T}} \cdot 13 \cdot \cancel{T}$$

On the right side, the T in the denominator cancels the T in the numerator.

This is the step to which cross-multiplying would have brought us directly. We continue solving as usual and calculate $93 \cdot 13$.

$$6T = 93 \cdot 13$$

$$6T = 1209$$

To solve for T, we need to divide both sides by 6.

$$\frac{6T}{6} = \frac{1209}{6}$$

On the left side, we simplify 6 and 6 so that only T is left. On the right side, we perform the division.

$$T = 201.5$$

This is the final answer.

Cross-multiplying is not a "magic trick" but simply a *shortcut* based on mathematical principles.

1. Solve the proportion in two ways: (a) using the shortcut of cross-multiplying,
   (b) using the slow way as in the example above.

a. $\dfrac{8}{1.15} = \dfrac{37}{K}$	**b.** $\dfrac{8}{1.15} = \dfrac{37}{K}$

Unit Rates

Remember that a rate is a ratio where the two terms have different units, such as 2 kg/$0.45 and 600 km/5 h.

In a **unit rate, the second term of the rate is one** (of some unit).

For example, 55 mi/1 hr and $4.95/1 lb are unit rates. The number "1" is nearly always omitted so those rates are usually written as 55 mi/hr and $4.95/lb.

To convert a rate into an equivalent unit rate simply divide the numbers in the rate.

Example 1. Mark can ride his bike 35 km in 1 ½ hours. What is the unit rate?	**Example 2.** A snail can slide through the mud 5 cm in 20 minutes. What is the unit rate?

Example 1. Mark can ride his bike 35 km in 1 ½ hours. What is the unit rate?

To find the unit rate, we use the principles of division by fractions to divide 35 km by 1 ½ h. The units "km" and "hours" are divided, too, and become "km per hour" or "km/hour."

$$\frac{35 \text{ km}}{1 \ 1/2 \text{ h}} = 35 \div \frac{3}{2} \text{ km/h} = 35 \times \frac{2}{3} \text{ km/h}$$

$$= \frac{70}{3} \text{ km/h} = 23 \tfrac{1}{3} \text{ km/h}.$$

We could also use decimal division:
35 km ÷ 1.5 h = 23.333... km/h.

So the unit rate is 23 ⅓ km per hour.

Example 2. A snail can slide through the mud 5 cm in 20 minutes. What is the unit rate?

Here, it is actually not clear whether we should give the unit rate as cm/min or cm/hr. Let's do both.

(1) To get the unit rate in cm/min, we simply divide 5 cm ÷ 20 min. We get the fraction 5/20. We also divide the units to get "cm/min." So we get

   5 cm ÷ 20 min = 5/20 cm/min = 1/4 cm/min

Or use decimals: 5 cm/20 min = 25/100 cm/min

   = 0.25 cm/min.

(2) For centimeters per hour, we multiply both terms of the rate by 3 to get an equivalent rate of 15 cm in 60 minutes, which is 15 cm in 1 hour.

1. Find the unit rate.

   a. $125 for 5 packages

   b. $6 for 30 envelopes

   c. $1.37 for ½ hour

   d. 2 ½ inches per 4 minutes

   e. 24 m^2 per 3/4 gallon

2. A person is walking 1/2 mile every 1/4 hour. Choose the correct fraction for the unit rate and simplify it.

$$\frac{\frac{1}{4}}{\frac{1}{2}} \text{ miles per hour} \quad \text{or} \quad \frac{\frac{1}{2}}{\frac{1}{4}} \text{ miles per hour}$$

3. Write the unit rate as a complex fraction, and then simplify it.

a. Lisa can make three skirts out of 5 ½ yards of material. Find the unit rate for one skirt.

b. A drink made with 30 g of powdered vegetables gives you 2 ¾ servings of vegetables. Find the unit rate for 1 g of powder.

c. Marsha walked 2 ¾ miles in 5/6 of an hour.

d. It takes Linda 2 ½ hours to make 1 ½ vases by hand.

e. There are 5,400 people living in a suburban development area of 3/8 km^2.

f. Alex paid $8.70 for 5/8 lb of nuts.

g. Elijah can finish 3/8 of a game in 7/12 of an hour.

Example 3. On rough country roads, Greg averages a speed of 30 kilometers per hour with his moped.

This 30 km/h is a rate that involves two quantities—kilometers and hours, or distance and time — which we can consider as variables. See the table.

distance (km)	0	30	60	90	120	150
time (hours)	0	1	2	3	4	5

These two variables d and t are related by the equation $d = 30t$ that we can plot on the coordinate grid:

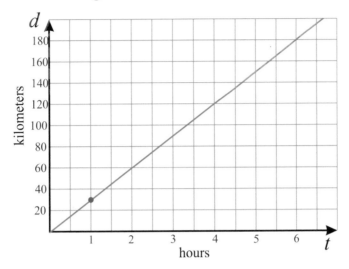

The unit rate 30 km/h is the slope of the line. It is also the coefficient of the variable t in the equation $d = 30t$.

We have plotted the point (1, 30) that matches the unit rate—1 hour and 30 kilometers.

What does the point (4, 120) mean?

It means that Greg can travel 120 kilometers in 4 hours.

4. A delivery truck is traveling at a constant speed of 50 km per hour.

   a. Write an equation relating the distance (d) and the time (t).

   b. Plot the equation you wrote in part (a) and the point that matches the unit rate.

   c. What does the point (3, 150) mean in terms of this situation?

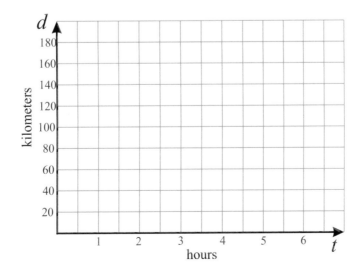

27

5. Some baby ducks are walking at a constant speed of 1/3 meter per second (or 1 meter in 3 seconds).

    a. Write an equation relating the distance (d) and time (t) and plot it in the grid below.

    b. What is the unit rate?

    c. Plot the point that matches the unit rate in this situation.

    d. What does the point (0, 0) mean in terms of this situation?

    e. Plot the point that matches the time $t = 4$ s.

    f. Plot the point that matches the distance $d = 3$ m.

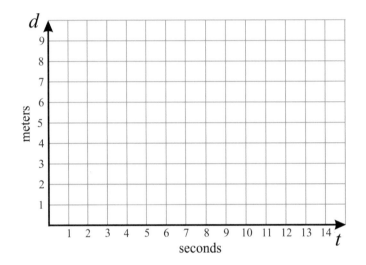

6. The equation $d = (1/2)t$ represents the distance in meters that adult ducks walk in t seconds.

    a. Plot this equation in the same grid as you did the equation for the baby ducks.

    b. Plot the point that matches the unit rate in this situation.

    c. How can you see from the graphs that the adult ducks walk faster than the babies?

    d. How much farther will the adult ducks have walked than the baby ducks at $t = 5$ s?

    e. How much longer will the baby ducks take to walk 5 meters than the adult ducks?

> **Example 4.** A town has 45,000 inhabitants and 128 doctors.
> (1) Find the number of doctors per 10,000 population.
> (2) Find the unit rate (the number of people per one doctor).
>
> (1) We can easily write the rate of doctors to all people — it is 128 doctors : 45,000 people. Since there are 4.5 groups of 10,000 people in 45,000, if we put 4.5 in place of the 45,000 in that rate and convert it to a unit rate, then we get the rate for one group of 10,000 people.
>
> So first we write the ratio 128 to 4.5, which is the ratio of doctors to 10,000 people. Then we divide $128 \div 4.5 = 28.444...$ to get the actual number of doctors for 10,000 people. So there are about 28 doctors for each 10,000 people.
>
> (2) We divide the number of people by the number of doctors: 45,000 people $\div$ 128 doctors $= 351.5625$ people/doctor ≈ 352 people/doctor. In other words, on average, each doctor serves about 352 people.

7. **a.** Calculate the rate of physicians per 10,000 people in Bulgaria, if the country is estimated to have 27,700 doctors and 7,365,000 people. Round your answer to one decimal.

   b. Algeria has 12.1 physicians per 10,000 people. How many doctors would you expect to find in an area in Algeria that has 350,000 residents?

   c. What is the rate of physicians per 1,000 people in Algeria?

8. Jane and Stacy ran for 30 seconds. Afterward each girl checked her heartbeat. Jane counted that her heart beat 38 times in 15 seconds, and Stacy counted that her heart beat 52 times in 20 seconds.

   a. Which girl had a faster heart rate?
   How much faster?

   b. Let's say Jane keeps running and her heart keeps beating at the same rate. Write an equation for the relationship between the number of her heartbeats and time in seconds.
   Also, identify the unit rate in this situation.

   c. Do the same for Stacy.

Proportional Relationships

In this lesson we study what it means when two variables are **in direct variation** or **in proportion**. The basic idea is that whenever one variable changes, the other varies (changes) **proportionally** or **at the same rate.**

Example 1. Apples cost $3.50 per kilogram.

As you already know, "$3.50 per kilogram" is a unit rate. It relates two quantities: the cost and the weight of the apples. We will now consider those quantities to be *variables* and let them vary.

In this situation the two variables—the <u>cost</u> and the <u>weight</u>—are in **direct variation** or **in proportion**. This means that if either variable doubles, to maintain the proportion, the other must also double. If one variable increases ten times, the other also has to increase ten times. If one of them is cut to a third, the other must also be cut to a third, and so on. If you multiply or divide either variable by any number, the other variable must get multiplied or divided by the same number.

At $3.50 per kilogram, 6 kg of apples would cost $21. If I double the weight to 12 kg, the cost also doubles, to $42. If I want only 1/3 of 6 kg, or 2 kg, the cost is only 1/3 of $21, or $7.

In summary, whenever one variable in the rate changes, the other has to change proportionally.

You can check to see if two variables are in direct variation in several different ways. Here is one way.

(1) Check to see if the values of the variables are in direct variation. If you double the value of one, does the value of the other double also? If one quantity increases by 5 times, does the other do the same?

>   **Example 2.** The table gives the cost of using a computer in an internet cafe based on the number of hours of usage. Are the cost and time in proportion?

Cost ($)	3	4	5	6	7	8	9
Time (hr)	1	2	3	4	5	6	7

Look at the two rates $4 for 2 hours and $8 for 6 hours. The time triples but the cost only doubles! Therefore, the relationship between these two quantities (cost and time) is not proportional.

1. Fill in the table of values and determine whether the two variables are in direct variation.

a. $y = 3x$

y								
x	−3	−2	−1	0	1	2	3	4

b. $y = x + 2$

y								
x	−3	−2	−1	0	1	2	3	4

c. $y = (1/2)x - 1$

y								
x	−3	−2	−1	0	1	2	3	4

d. $y = -x$

y								
x	−3	−2	−1	0	1	2	3	4

e. $C = 2.4n$

C								
n	0	1	2	3	4	5	6	7

f. $h = 1/k$

h								
k	1	2	3	4	5	6	7	8

There is a basic difference between the graphs of equations that have variables in proportion and those that do not. There is also a difference in the equations themselves. Try to spot those differences in the following exercise.

2. Fill in the table of values and determine whether the two variables are in direct variation. Then plot the equations.

a. $y = x + 1$

x	−3	−2	−1	0	1	2	3	4
y								

b. $y = 2x$

x	−3	−2	−1	0	1	2	3	4
y								

c. $y = 2x − 1$

x	−3	−2	−1	0	1	2	3	4
y								

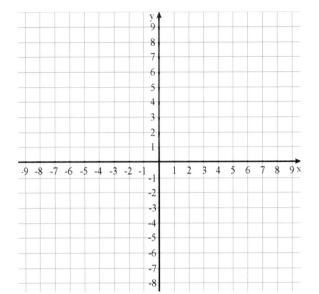

d. $y = (1/2)x$

x	−3	−2	−1	0	1	2	3	4
y								

e. $y = −2x$

x	−3	−2	−1	0	1	2	3	4
y								

f. $y = −2x + 1$

x	−3	−2	−1	0	1	2	3	4
y								

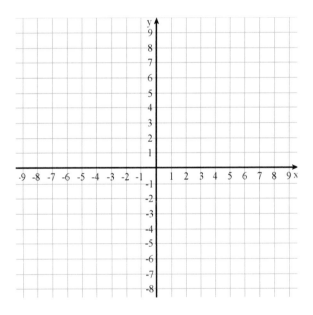

3. Now consider the plots, the equations, and the tables of values of the six items in the previous exercise. How do the equations and plots of the variables that are in proportion differ from those that aren't? If you cannot tell, check the next page.

(2) When two variables are proportional, the equation relating the two is of the form $y = mx$**, where** y and x **are the variables, and** m **is a constant. The constant** m **is called the constant of proportionality.**

For example, if apples cost \$3.50 per kilogram, the equation relating the weight (w) to the cost (C) of the apples is $C = 3.5w$. The constant of proportionality is 3.5.

- The constant of proportionality is also the unit rate. When the weight of apples is 1 kg, the cost is $C = \$3.5 \cdot 1 = \$3.50$.

(3) When two quantities are proportional, their graph is a straight line that goes through the origin.

Remember, the equation relating the two quantities is of the form $y = mx$. The constant m is not only the unit rate—it is also the slope. In our apple example, the equation is $C = 3.5w$. The slope is 3.5, because whenever the weight (w-value) increases by 1 kg, the cost (C-value) increases by \$3.50.

Notice the point (1, 3.5) that corresponds to the unit rate: the weight is 1 kg and the cost is \$3.50.

The other special point on the graph is the origin at (0, 0). That point is always on the graph because, if two quantities are in proportion, when one of them is zero, the other also has to be zero.

We graph the weight w on the horizontal axis as the independent variable and the cost C on the vertical axis as the dependent variable because we are choosing a weight of apples independently based on some need and then "observing" what its cost is. So here the cost C depends on the weight w. The dependent variable depends on the independent variable. We always plot the independent variable on the horizontal axis.

Cost of Apples

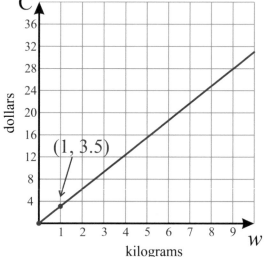

It is possible to look at the situation just the opposite way, and consider how the weight of the apples depends on the cost of the apples. In that case, we would write the equation $w = 0.2857C$ and plot the values of C on the horizontal axis. This way is not as common as observing how the cost depends on the weight.

Just one more thought: You might wonder, "Why does the line have to go through the origin if the quantities are in proportion?"

Consider the principle governing direct variation: if one quantity is cut in half, then the other is cut in half. Let's say you start with certain values of the two quantities, such as 6 meters per 2 minutes. Now cut both in half to get 3 meters per 1 minute. Do it again to get 1.5 meters per 1/2 minute. Do it again, and again.

Notice that both numbers get smaller and smaller yet—they approach zero. This would happen no matter what values of the two quantities you started with. So the point (0, 0) has to be included in the graph of quantities that are in direct variation.

4. **a.** Graph the equation $y = 3x$.

   b. State the unit rate in this situation.

   c. Plot the point that corresponds to the unit rate.

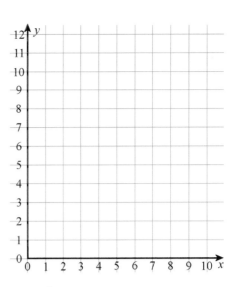

5. **a.** Graph the equation $y = 0.5x$.

   b. State the unit rate in this situation.

   c. Plot the point that corresponds to the unit rate.

6. Determine whether the two quantities are in proportion. If so, find the unit rate, write
   an equation relating the two, and graph the equation.

a.

x	−3	−2	−1	0	1	2	3	4
y	−5	−4	−3	−2	−1	0	1	2

In proportion or not?

Unit rate:

Equation:

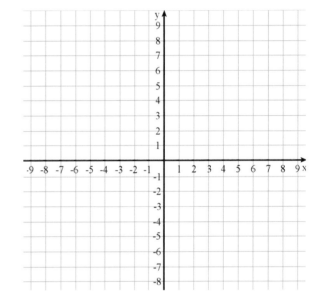

b.

x	−3	−2	−1	0	1	2	3	4
y	−12	−8	−4	0	4	8	12	16

In proportion or not?

Unit rate:

Equation:

c.

x	−3	−2	−1	0	1	2	3	4
y	−1	−2/3	−1/3	0	1/3	2/3	1	4/3

In proportion or not?

Unit rate:

Equation:

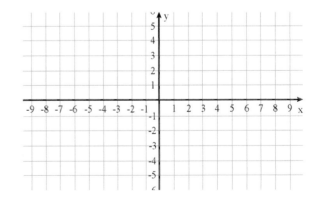

d.

x	−3	−2	−1	0	1	2	3	4
y	2	5/3	4/3	1	2/3	1/3	0	−1/3

In proportion or not?

Unit rate:

Equation:

7. For each line:
   (1) state the unit rate (including the units of measurement),
   (2) plot the point that corresponds to the unit rate, and
   (3) write an equation for the line.

   a. Unit rate:

   Equation:

   b. Unit rate:

   Equation:

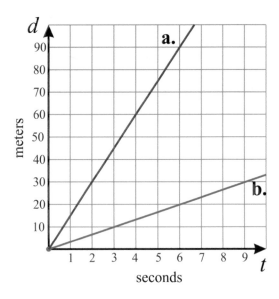

8. Write an equation for each line.

   a. Equation:

   b. Equation:

   c. What real-life situation could the equations
   and their plots represent?

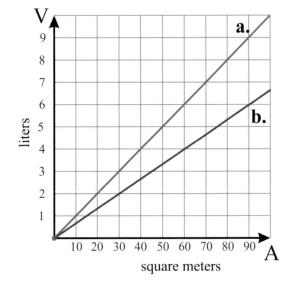

9. A car is traveling at a constant speed, covering
   135 miles in three hours.

   a. Write an equation that relates the distance
   the car has traveled to the time it takes
   to do so.

   b. Plot the equation on the grid. Choose the
   scaling on the axes so that you can fit the
   point $t = 10$ hours and the corresponding
   distance onto the grid.

   c. Plot the point that corresponds to the unit rate.

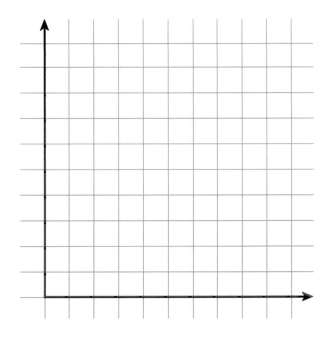

10. Robert works as a salesman. He is paid $300 per week plus a commission based on the total sales he makes. Consider Robert's pay and the number of items he sells.

Items sold in a week	0	1	2	3	4	5	6	7	8
Pay	300	350	400	450	500	550	600	650	700

    a. Are these two quantities in proportion?

    b. If so, write an equation relating the two and state the constant of proportionality.

11. Three workers planting trees on a big farm can plant 60 trees in a day, on average. Consider the number of workers and the number of trees they plant in one day.

    a. Are these two quantities in proportion?

    b. If so, write an equation relating the two and state the constant of proportionality.

12. The table shows the relationship between the number of workers and the time it takes to finish painting a house.

Number of workers	1	2	3	4	5	6
Time (hours)	10	5	3.3	2.5	2	1.7

    a. Are these two quantities in proportion?

    b. If so, write an equation relating the two and state the constant of proportionality.

13. The monthly cost for a cell phone depends on the total time the phone is used for calls, as shown in the graph.

    a. Are the two quantities, the cost C and the calling time t, in proportion?

    b. What are the coordinates of the marked point?

    c. What does that point mean in terms of this situation?

Cost of using a cell phone

Graphing Proportional Relationships – More Practice

You may use a calculator for all the problems in this lesson.

1. Shelly uses up a 400-ml bottle of shampoo in 7 months.

    a. What is the unit rate in this situation?

    b. Write an equation relating the amount of shampoo (S) in milliliters to the time (t) in months.

    c. Plot your equation. Choose appropriate scaling for the two axes.

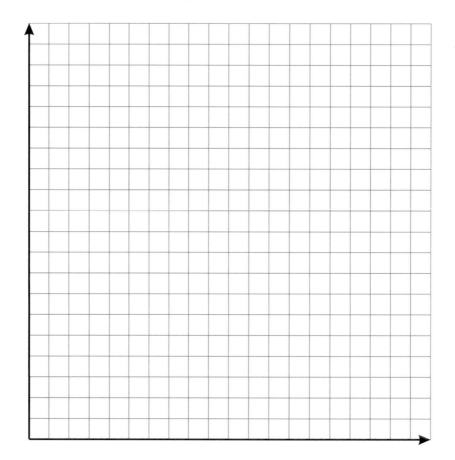

    d. Plot a point that corresponds to the time t = 3 months.

    e. Plot a point that corresponds to the unit rate.

    f. Using shampoo at the same rate, how long would it take her to use 250 ml of shampoo?

2. Sarah works as a secretary at a hospital. She gets paid $100 for an 8-hour workday.

   a. Write an equation that relates Sarah's pay to the time (in hours) that she has worked.

   b. Plot the equation on the grid. Choose the scaling on the axes so that you can fit the point that corresponds to *time* = 20 hours onto the grid.

   c. Plot the point that corresponds to the unit rate.

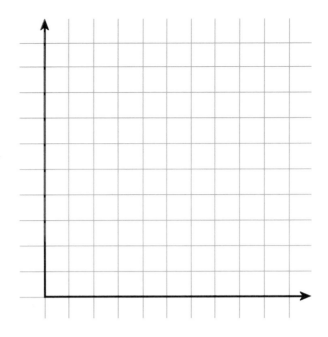

3. A car can travel 28 miles on a gallon of gasoline, and gasoline costs $3 a gallon. Notice that this situation actually involves *three* quantities: the mileage, the amount of gasoline used, and the cost of gasoline.

   a. We will now consider the mileage (m) and the amount of gasoline (g). Write an equation that gives you the mileage in terms of the gasoline used (in the form m = (expression)).

   b. Plot your equation. Plot the independent variable on the horizontal axis. Make sure you can fit 20 gallons on the g-axis and the distance the car travels on 20 gallons of gasoline on the m-axis.

   c. Plot a point that corresponds (approximately) to $m = 100$ miles.

   d. Plot a point that corresponds to $g = 10$ gallons.

   e. Plot a point that corresponds to the unit rate.

   f. What does the point $(0, 0)$ mean in this situation?

   g. How far could the car travel on 100 gallons of gasoline?

   Now we will also consider the third quantity, the *cost*.

   h. How far could the car travel on $100 of gasoline?

   i. Find the cost of traveling 700 miles.

More on Proportions

You can often solve a rate problem in several different ways. Try to find more than one way to solve the problems on this page. You can get some ideas from the example below.

Example 1. Tim can type 350 words in 9 minutes. Typing at the same speed, how many words could he type in 40 minutes?

Solution with a proportion:

$$\frac{350 \text{ words}}{9 \text{ min}} = \frac{w}{40 \text{ min}}$$

$$9 \text{ min} \cdot w = 350 \text{ words} \cdot 40 \text{ min}$$

$$\frac{9 \text{ min} \cdot w}{9 \text{ min}} = \frac{14{,}000 \text{ words} \cdot \text{min}}{9 \text{ min}}$$

$$w \approx 1556 \text{ words}$$

Solution with a unit rate:

The rate of 350 words per 9 minutes means $350 \div 9 = 38.\overline{8}$ words per minute.

In 40 minutes he can type 40 times that many words, or $40 \text{ min} \cdot 38.\overline{8}$ words per minute $= 1{,}555.\overline{5}$ words ≈ 1556 words.

Solution in a diagram form:

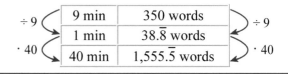

You may use a calculator for all the problems in this lesson. Choose one (or more) of the exercises 1-3 and try to find several ways to solve it.

1. The total weight of 18 identical books is 27.4 pounds.
   How much do five of those books weigh?
   Try to find more than one way to solve this problem.

2. Fencing costs $45 for a roll of 8 meters, but the vendor will also let you purchase part of a roll.
   So how much will it cost to fence a rectangular 15 m × 20 m plot?
   Try to find more than one way to solve this problem.

3. If a boy can ride his bike 15 km in 37 minutes, how far could he ride in one hour (at the same speed)?
   Try to find more than one way to solve this problem.

Comparing rates

Once again, there are several ways to determine if two rates are equal or which of two rates is greater.

First try to solve the exercises 4 and 5 on your own! Then read the example at the bottom of the page.

4. One particular pasta sauce costs $3.95 for 450 g and another costs $4.55 for 560 g.
   Are the two rates equal? If not, which sauce costs more per gram?

5. Are these two rates equal? 50 miles / 2.2 gallons and 125 miles / 5.5 gallons.
   If not, which one is greater?

Example 2. John applied 10 kg of fertilizer to a field of 600 m^2, and the next day he applied 16 kg of fertilizer to a field of 1500 m^2. Are the rates equal?

(1) Solution using a proportion:

If the rates are equal, they are in proportion, and cross-multiplying will produce a true equation.

$$\frac{10 \text{ kg}}{600 \text{ m}^2} \overset{?}{=} \frac{16 \text{ kg}}{1500 \text{ m}^2}$$

$$10 \text{ kg} \cdot 1500 \text{ m}^2 \overset{?}{=} 16 \text{ kg} \cdot 600 \text{ m}^2$$

$$15{,}000 \text{ kg} \cdot \text{m}^2 \neq 9{,}600 \text{ kg} \cdot \text{m}^2$$

We didn't get a true equation, so the rates are not equal.

(2) Solution using the unit rates:

The first unit rate is 10 kg/600 m^2 = 0.01$\overline{6}$ kg/m^2.

The second is 16 kg/1500 m^2 = 0.010$\overline{6}$ kg/m^2.

The unit rates are not equal, so the original rates are not equal either. The first unit rate (and thus the first rate) is actually quite a bit more than the second.

(3) Solution with logical reasoning

If 10 kg was enough for 600 m^2, and the next day he used 1.6 times that much (16 kg), then the area should be 1.6 times 600 m^2. However, 1500 m^2 is 2 1/2 times 600 m^2. So the rates are not equal.

6. Write a word problem that can be solved by the given proportion in the lower box. Then solve the proportion. Round your answer to a meaningful accuracy. Don't forget to check that your answer is reasonable.

a. Word problem:	**b.** Word problem:
a. Proportion: $$\dfrac{60 \text{ words}}{97 \text{ sec}} = \dfrac{10{,}000 \text{ words}}{x}$$	**b.** Proportion: $$\dfrac{\$5.59}{80.0 \text{ g}} = \dfrac{\$100}{x}$$

7. The cost of electricity to run an air conditioner for 24 hours is $3.60. How much would it cost to run the air conditioner for a full week if it is on only 14 hours a day?

8. A man can carry 90 lb. If ten identical books weigh 27 lb, then how many of those books can the man carry?

9. Elijah needs to apply fertilizer on the lawn at the golf club. The instructions say to apply 1 pound of nitrogen per each 1,000 square feet of lawn. The fertilizer he uses comes in 10-lb bags and is composed of 25% nitrogen and 75% other minerals. How many 10-lb bags of fertilizer does Elijah need for a rectangular lawn of 300 ft by 1,200 ft?

Let's say that the ratio of quantities y and x is always 2/3. The table below shows some possible values of x and y.

x	1	2	3	4	5
y	1.5	3	4.5	6	7.5

Since the ratio x/y is always 2/3, we can write a proportion: $\dfrac{x}{y} = \dfrac{2}{3}$.

You have learned that if two quantities are in proportion, the equation relating them is in the form $y = mx$. Can the proportion above be written in the form of $y = mx$ for some constant m? If so, what is the value of m?

Scaling Figures

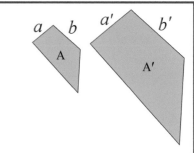

Two figures are **similar** if they have the same shape. Similar figures may be of different sizes. For example, all circles are similar, and so are all squares.

Example 1. The quadrilaterals A and A′ (read: "A prime") at the right are similar: they have the same basic shape, but one is larger.

Compare the corresponding sides: a to $a′$ and b to $b′$. In the case of polygons, similarity means that **corresponding sides are proportional** (in the same ratio) and corresponding angles are equal.

So the ratio $a : a′$ is equal to the ratio $b : b′$. This ratio is called the **similarity ratio**.

Example 2. The similarity ratio between the two rectangles is 2:7. Find the length of the side marked x.

Solution 1. The lengths of the corresponding sides are in the ratio of 2:7. The unknown length of the side x corresponds to the 2 parts of the ratio and the known 3.5 m side corresponds to the 7 parts. So each part is 3.5 m ÷ 7 = 0.5 m. The unknown length is 2 · 0.5 m = 1 m.

That makes sense, since we would expect the side marked with x to be quite a bit shorter than 3.5 m.

Solution 2. We write the ratio of the lengths of the corresponding sides and set that ratio to be 2/7. We get an equation involving two equal ratios—a proportion. Its solution is on the right.

$$\frac{x}{3.5 \text{ m}} = \frac{2}{7}$$
$$7x = 2 \cdot 3.5 \text{ m}$$
$$7x = 7 \text{ m}$$
$$x = 1 \text{ m}$$

1. The figures are similar. Find the length of the side labeled x.

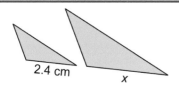

a. Similarity ratio 3:5.

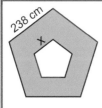

b. Similarity ratio 7:3.

2. The sides of two similar triangles are in a ratio of 3:4. If the sides of the larger triangle are 4.8 cm, 6.0 cm, and 3.6 cm, what are the sides of the smaller triangle?

3. Draw any triangle on blank paper or below. Then draw another, bigger triangle using the similarity ratio 2:5. Remember that corresponding angles in the two triangles will be equal.

4. The rectangles 1, 2, 3, and 4 in the table are similar.

   a. Consider the columns for length and width only. Complete the table by filling in the missing widths.

   b. In the last column, write the *aspect ratio* (the ratio of length to width) of each rectangle in simplified form. For example, for Rectangle 4, the aspect ratio is 2.5 cm : 7.5 cm = 25 : 75 = 1 : 3. What do you notice?

	Length	Width	Aspect Ratio
Rectangle 1	1 cm		
Rectangle 2	1.5 cm		
Rectangle 3	2 cm		
Rectangle 4	2.5 cm	7.5 cm	

Example 2. Sometimes you need to look carefully to find the corresponding sides. The two rectangles at the right are similar. Notice that the "top" sides of 18 ft and 22 ft do *not* correspond. Instead, the 18 ft side corresponds to the 10 ft side, because they are the *shorter* sides of the rectangles.

Here, the similarity ratio is 18 ft : 10 ft = 9:5. Also, since the side x units long corresponds to the 22 ft side, the ratio x : 22 ft is equal to ratio 9:5.

18 ft

22 ft

x   10 ft

5. The figures are similar. Find the length of the side labeled with x.

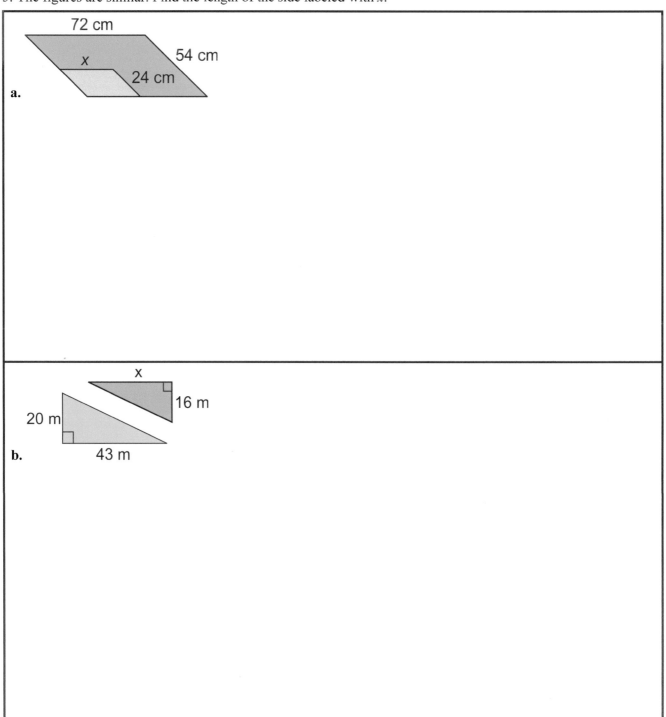

a.

b.

Scaling means to enlarge or shrink a figure while maintaining its shape. The resulting figure is therefore similar to the original one. The number by which all distances or dimensions of the figure are multiplied is called the **scale factor** or just the **scale**.

Example 3. A square was enlarged. What is the scale factor?

Since the 48 cm-side became 72 cm long, the scale factor from the smaller square to the larger one is 72/48 = 3/2 = 1.5. So each side of the square became 1.5 times as long as before.

48 cm   72 cm

We can also consider the **scale ratio** (or **ratio of magnification**): it is the ratio of any side of the figure after scaling it to its corresponding side before scaling (*after : before*). This ratio is the same, no matter which side is chosen. In this case, the scale ratio is 72:48 = 3:2. (Note that we don't write it as 48:72 because 48:72 is less than 1 and that would mean the figure shrank.)

The scale factor and the scale ratio are equal. In this case, the scale factor was 3/2 or 1.5, and the scale ratio was 3:2. Simply put, you get the scale factor by writing the scale ratio as a fraction or as a decimal.

Here is a way to keep the two very similar-sounding terms straight: The scale *ratio* is a ratio of two numbers (like 3:1), but the scale *factor* is a single number (such as 3).

6. **a.** Find the scale *factor* from the smaller to the larger parallelogram.

6 cm   19 cm   14 cm

   b. What is the scale *ratio*?

7. A rectangle with 20 ft and 48 ft sides is shrunk so that its sides become 15 ft and 36 ft.

   a. What is the scale ratio?

   b. What is the scale factor?

8. The area of a square is 36 cm². The square is shrunk using the scale factor 3/4. What is the area of the resulting square?

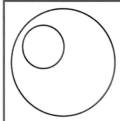

Example 4. A circle was shrunk. Find the scale ratio and the scale factor.

Any corresponding dimension — the diameter, the radius, or the circumference — will work for finding the scale ratio and scale factor. Measuring the diameters is the easiest.

The diameters measure 3.5 cm and 1.4 cm. Since the circle was shrunk, the scale ratio and the scale factor are less than 1, so we write the ratio as 1.4 : 3.5 and not vice versa.

However, it is customary to express ratios using whole numbers when possible. Multiplying both terms of the ratio by 10 gives us an equivalent ratio without decimals that we can reduce to lowest terms:

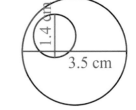

$$\frac{1.4}{3.5} = \frac{14}{35} = \frac{2}{5}$$

Writing 2/5 as a decimal, we get the scale factor 0.4. So the dimensions of the smaller circle are 2/5 or 0.4 times the dimensions of the larger one.

9. The triangle on the left was scaled to become the triangle on the right. Find the scale ratio, and then write it as a scale factor. Use a ruler that measures in centimeters.

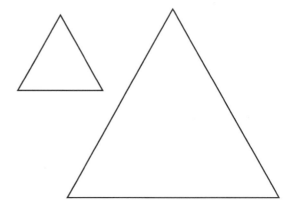

10. Enlarge this L-shape using the scale ratio 3:2. Draw the resulting larger L-shape beside it.

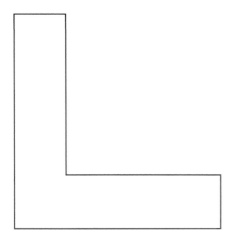

11. The sides of a rectangle measure 3″ and 4 1/2″. The shorter side of another, similar rectangle is 3/4″.

a. In what ratio are the sides of the two rectangles?

b. What is the length of the other side of the similar rectangle?

c. Calculate the areas of both rectangles.

d. In what ratio are their areas?

Puzzle Corner   The aspect ratio of a rectangle is 2:3 and its perimeter is 50 cm.
The rectangle is shrunk in a scale ratio of 2:5. What is its area now?

Floor Plans

Floor plans are drawn using a **scale**, which is a ratio relating the distances in the plan to the distances in reality. For example, a scale of 1 cm : 2 m means that 1 cm in the drawing corresponds to 2 m in actual size.

Example 1. In reality, how big is a room that measures 1 ¾″ by 2 ½″ in a plan with a scale of 1 in : 10 ft?

Since 1 inch corresponds to 10 ft, we simply need to multiply the length and width given in inches by 10.

Using decimals, the dimensions are 1.75 in by 2.5 in. So the dimensions of the room in actual size are $1.75 \cdot 10 = 17.5$ ft and $2.5 \cdot 10 = 25$ ft.

However, we're really not just multiplying by the number 10 but by the ratio 10 ft/1 in. That's how we keep track of the units to make sure that our final answer ends up with the correct units (feet and not inches). This is what's really happening in the calculation:

$$1.75 \; \cancel{\text{in.}} \cdot \frac{10 \text{ ft}}{1 \; \cancel{\text{in.}}} = 17.5 \text{ ft} \quad \text{and} \quad 2.5 \; \cancel{\text{in.}} \cdot \frac{10 \text{ ft}}{1 \; \cancel{\text{in.}}} = 25 \text{ ft}$$

Why not multiply by 1 in/10 ft (or 1 in : 10 ft) as the ratio is stated in the problem? Then the inches ("in") in the dimension wouldn't cancel the inches in the conversion factor, and we would end up "in²/ft" as our unit of length, instead of "ft."

Example 2.

A room measures 4.2 m by 3.2 m. How big would it appear on a floor plan with a scale 1 cm : 0.8 m?

Instead of multiplying by the scale ratio like we did in Example 1, we can set up one proportion for the length (L) and another for the width (W) and solve them. The length is 5.25 cm, and the width is 4 cm.

$\dfrac{1 \text{ cm}}{0.8 \text{ m}}$	$=$	$\dfrac{L}{4.2 \text{ m}}$
$0.8L$	$=$	4.2 cm
$\dfrac{0.8L}{0.8}$	$=$	$\dfrac{4.2 \text{ cm}}{0.8}$
L	$=$	5.25 cm

$\dfrac{1 \text{ cm}}{0.8 \text{ m}}$	$=$	$\dfrac{W}{3.2 \text{ m}}$
$0.8W$	$=$	3.2 cm
$\dfrac{0.8W}{0.8}$	$=$	$\dfrac{3.2 \text{ cm}}{0.8}$
W	$=$	4 cm

1. This room is drawn at a scale of 1 in : 4 ft. Measure dimensions asked below from the picture and then calculate the actual (real) dimensions.

   a. the bed

   b. the desk

2. What is the area of this room in reality?

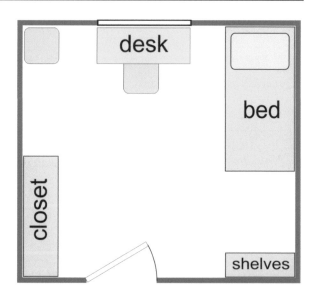

1 in : 4 ft

3. In the middle of the plan for the room, draw a table that in reality measure 3.5 ft × 2.5 ft.

4. A room measures 4 ½ inches by 4 inches in a plan
   with a scale of 1 in : 3 ft.

   a. What would the dimensions of the room be if it was
   drawn to the scale of 1 in : 6 ft?

   b. What would the dimensions of the room be if it was
   drawn to the scale of 1 in : 4 ft?

5. This room measures 2.5 m by 3.5 m in reality.

   a. To what scale is it drawn here?

   b. On the plan, to scale, draw:

   - two windows that are 80 cm wide,

   - a door that is 1 m wide, and

   - a bed that is 150 cm by 200 cm.

6. A floor plan is drawn using the scale 5 cm : 1 m.

   a. Calculate the dimensions in the plan for a kitchen
   that measures 4.5 m by 3.8 m in reality.

   b. The living room measures 26 cm by 22.5 cm on the plan.
   What are its dimensions in reality?

7. In the space at the right, draw a plan for
   a room that measures 3.8 m by 4.6 m. Put
   a 120 cm by 120 cm table in the middle
   of the room. Use the scale 2 cm : 1 m.

8. This is a floor plan for a small cottage at the scale 1/8 in : 1 ft. We have omitted the doors and windows to
   keep it very simple. (You can add some if you would like.)

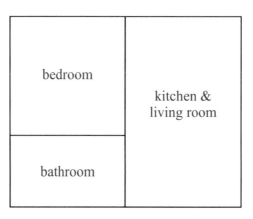

Scale 1/8 in : 1 ft

   a. What is the area of the cottage in reality?

   b. Redraw the plan at the scale 3/8 in : 1 ft on blank paper.

9. Redraw this floor plan at the scale 1 cm : 125 cm.

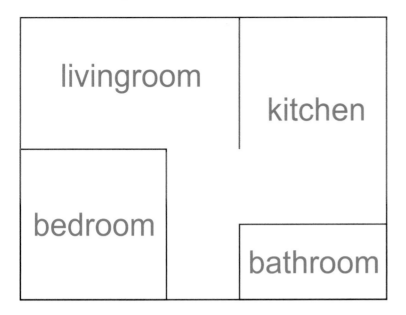

Scale 1 cm : 1 m

Maps

Just like floor plans, maps also include a scale. A scale on a map may show how units on the map correspond to units in reality (for example 1 cm = 50 km). It can also be given as a ratio such as 1:120,000.

A scale of 1:120,000 means that 1 unit on the map corresponds to 120,000 units in reality. This holds true—whether you use centimeters, millimeters, or inches—because the scale 1:120,000 is a ratio without any particular unit. So 1 cm on the map corresponds to 120,000 cm in reality, and 1 inch on the map corresponds to 120,000 inches in reality.

Example 1. A map has a scale 1:150,000. How long in reality is a distance of 7.1 cm on the map?

Below you can read two solutions to this problem. Both are actually very similar!

Multiply, then change the units.

If 1 cm corresponds to 150,000 cm, then 7.1 cm corresponds to 7.1 · 150,000 cm = 1,065,000 cm.

To be useful, this figure needs to be converted into kilometers. You can do this in two steps:

1. From centimeters to meters: Since 1 m = 100 cm, we remove two zeros from 1,065,000 cm to get 10,650 meters (or you can think of it as dividing by 100).

2. From meters to kilometers: Since 1 km = 1,000 m, the 10,650 meters corresponds to 10.65 km ≈ 11 km.

Change the units, then multiply.

In this solution, we will first rewrite the scale and then use multiplication to calculate the distance in reality.

Since 1 cm corresponds to 150,000 cm, and 150,000 cm = 1,500 m = 1.5 km, we can rewrite the scale of this map as 1 cm = 1.5 km.

Then, 7.1 cm corresponds to 7.1 · 1.5 km = 10.65 km ≈ 11 km.

You can use a calculator for all the problems in this lesson.

1. A map has a scale ratio of 1:20,000. Fill in the table.

on map (cm)	in reality (cm)	in reality (m)	in reality (km)
1 cm	20,000 cm		
3 cm			
5.2 cm			
0.8 cm			
17.1 cm			

2. A map has a scale of 1:100,000.

    a. The scale says that 1 cm on the map corresponds to 100,000 cm in reality. How many kilometers is that?

    Thus, we can rewrite this scale in the format 1 cm = _____ km

    b. A ski trail measures 5.2 cm on this map. In reality, how long is the trail in kilometers?

3. A map has a scale of 1:25,000.

   a. Rewrite this scale in the format 1 cm = _____ m.

   b. Fill in the table. Give your answers to the nearest tenth of a centimeter.

on the map (cm)	in reality
	500 m
	900 m
	1.6 km
	2.5 km

4. Measure the aerial distances between the given places in centimeters and then calculate the distances in reality to the nearest kilometer. The places are marked with squares on the map. *Aerial distances* are "as the crow flies": measure them directly from point to point, not by following the roads.

   a. From Elkmont to the the Gatlinburg Welcome Center.

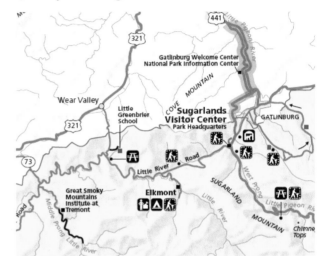

Scale 1:180,000

   b. From the Great Smoky Mountains Institute at Tremont to the Little Greenbrier School.

   c. From the Little Greenbrier School to Elkmont.

5. Hannah is making a map of a farmhouse and the surrounding buildings. When she measured the distance from the farmhouse to the barn, it was 75 meters.

   a. On the map, what will the distance be from the house to the barn at a scale of 1:500?

   b. What would the distance be on a map at a scale of 1:1200?

Example 2. The distance from Jane's home to her grandma's is 220 km.
How long is the representation of that distance on a map with a scale of 1:1,500,000?

Study both solutions below, and make sure you understand them.

Divide, then convert.

This conversion goes the other way: from reality to the map. Therefore, we need to *divide* the distance 220 km by the factor 1,500,000. We will get a very small number, and it is in kilometers just like the original distance is:

$$220 \text{ km} \div 1,500,000 = 0.000146\overline{6} \text{ km}$$

However, the answer in this format is not very useful. We need to convert it into units that can be measured on a map, such as centimeters (millimeters would work, too). You could convert $0.000146\overline{6}$ km into centimeters directly, but here we will do it in two steps because that is easier for most people.

(1) Converting from kilometers to meters requires multiplying our number by the unit ratio 1,000 m/km (not by 1 km/1000 m because then the units "km" wouldn't cancel):

$$0.000146\overline{6} \text{ k̶m̶} \cdot \frac{1,000 \text{ m}}{\text{k̶m̶}} = 0.146\overline{6} \text{ m}$$

(2) Finally, we convert meters into centimeters by multiplying by 100 cm/m. We get

$$0.146\overline{6} \text{ m̶} \cdot \frac{100 \text{ cm}}{\text{m̶}} = 14.6\overline{6} \text{ cm}.$$

So the distance on the map is about 14.7 cm.

Convert, then divide.

We will first rewrite the scale and then use division to calculate the distance in reality.

According to the scale, 1 cm corresponds to 1,500,000 cm. Now, 1,500,000 cm = 15,000 m (think of dropping two zeros) and that equals 15 km (think of dropping three zeros).

So we can rewrite the scale of this map as 1 cm = 15 km.

Thus 220 km corresponds to
$$220 \text{ km}/15 \text{ km} = 14.\overline{6} \approx 14.7 \text{ cm}.$$

6. Mark is planning a route for a footrace that will be 1.5 km long. He has two city maps available. One has a scale of 1:15,000 and the other has a scale of 1:20,000. Calculate the distance of the race on each of the two maps to the nearest tenth of a centimeter.

Example 3. A map has a scale of 1:500,000. The distance from one town to another measures 5 1/4 inches on the map. How long is the distance in reality?

Again, there are two ways to solve this problem:

Multiply, then convert: Multiply the given distance 5.25 in by 500,000 and then convert the result into miles.

Convert, then multiply: Rewrite the scale into an "easier" format, then use multiplication.

Multiply, then convert.

We simply multiply 5.25 in by 500,000 to get 5.25 in · 500,000 = 2,625,000 inches. So that is the distance in reality. Next we convert this distance into miles in two steps:

(1) <u>From inches to feet:</u> Since 1 ft = 12 in, we multiply 2,625,000 inches by the ratio 1 ft/12 in (not by 12 in/1 ft because we want the inches to cancel):

$$2{,}625{,}000 \text{ in.} \cdot \frac{1 \text{ ft}}{12 \text{ in.}} = 218{,}750 \text{ ft}$$

You end up dividing 2,625,000 by 12, which makes sense since we are going from smaller units (inches) to bigger ones (feet), and thus we should get *fewer* units in feet.

(2) <u>From feet to miles:</u> 1 mi = 5,280 ft. Again, we multiply by the conversion ratio 1 mi/5,280 ft:

$$218{,}750 \text{ ft} \cdot \frac{1 \text{ mi}}{5{,}280 \text{ ft}} = 41.4299\overline{24} \text{ mi} \approx 41 \text{ miles.}$$

(Essentially, you divide by 5,280.)

The distance between the towns is about 41 miles.

Convert, then multiply.

The scale 1:500,000 means that 1 inch corresponds to 500,000 inches. We need to convert that to a more useful unit, such as feet or miles. Similar to above, the conversions go like this:

$$500{,}000 \text{ in.} \cdot \frac{1 \text{ ft}}{12 \text{ in.}} = 41{,}666.\overline{6} \text{ ft} \quad \text{and} \quad 41{,}666.\overline{6} \text{ ft} \cdot \frac{1 \text{ mi}}{5{,}280 \text{ ft}} = 7.89\overline{14} \text{ mi} \approx 7.90 \text{ miles}$$

So the scale of our map is 1 inch = 7.90 miles.

Then, the given distance 5.25 miles corresponds to 5.25 · 7.90 miles ≈ 41 miles.

Either way, the most difficult part is in converting inches into miles. Since we are dealing with customary units, there aren't any shortcuts that would allow us to convert the measuring units without a calculator, so neither solution is really easier than the other.

---

7. A map has a scale ratio of 1:400,000. In miles, how long is a nature hike that measures 2.5 inches on the map? Give your answer to the nearest mile.

8. Use a map you have on hand, and measure distances on it with a ruler. Then calculate the distances in reality and give them to a reasonable accuracy. If you don't have a map on hand, skip this exercise and just do the next one.

9. On this map of the USA measure the distances in inches. Then calculate the distances in reality and give them to the nearest hundred miles.

   a. From Tallahassee to Denver.

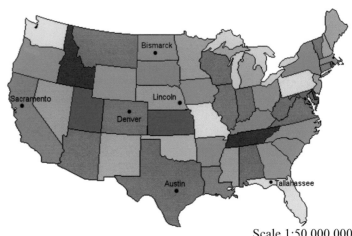

Scale 1:50,000,000

   b. From Sacramento to Austin.

   c. From Lincoln to Bismarck.

10. An island is 16.2 miles from the mainland. What is that distance on a map with a scale of 1:500,000? Finish Ellie's solution to this problem by filling in the words *multiply* and *divide*, the sign "·" or "÷," and numbers. Give your final answer to the tenth of a inch.

First, I _____ the distance 16.2 miles by the factor 500,000. I will get a very

small number, which will be in miles: 16.2 miles ▢ 500,000 = _____ miles.

Next I convert this to feet, and then to inches.

Converting miles to feet means to _____ by the ratio 5,280 ft/1 mi:

_____ · $\dfrac{5{,}280\text{ ft}}{1\text{ mi}}$ = _____

Then I convert the result from feet to inches by _____ing by the ratio 12 in/ 1 ft:

_____ · $\dfrac{12\text{ in.}}{1\text{ ft}}$ = _____ in. ≈ _____ in.

11. The distance from Mark's home to the airport is 45.62 miles according to an online distance calculator. How long would this distance be, in inches, on a map with a scale of 1:250,000? How about on a map with a scale of 1:300,000?

12. The scale of a map is 1:15,000. A rectangular plot of land measures 1 3/16″ by 2 1/8″ on the map.
    a. Find the area of the land in reality in square feet. Don't round your answer, as we will use the answer in part (b).

    b. Calculate the area of the land in acres, to the nearest tenth of an acre. Use 1 acre = 43,560 square feet.

13. The length of a hiking path is 5.0 inches on a map with a scale of 1:200,000. What would the length be on a map with a scale of 1:150,000?

A sheet of A4 paper measures 210 mm by 297 mm. You want to print a map of a plot of land with the dimensions of 1.65 km by 2.42 km onto one sheet of A4 paper. What scale should you use for your map so that it fits onto the sheet of A4 paper?

Puzzle Corner

Significant Digits
(This lesson is optional.)

Example 1. In reality, how long is a distance of 5.7 cm on a map with a scale of 1:400,000?

Since 1 cm corresponds to 400,000 cm, then 5.7 cm corresponds to 5.7 · 400,000 cm = 2,280,000 cm. Converting this into kilometers we get 22.8 km.

However, since our measurement was only to the accuracy of a tenth of a centimeter, we cannot truthfully give our answer to an accuracy of 22.8 km. You see, the measurement 5.7 cm is an *approximation*. The true distance on the map could be 5.7352 cm or 5.67364 cm — we don't know since we cannot measure it that accurately.

Let's consider some other distances on the map that would be rounded to 5.7 cm, and calculate them in reality. Study the table on the right:

From the table we can see that the distance in reality is anywhere from 22.6 km to about 23 km. We definitely cannot say it is exactly 22.8 km. That is why we need to round 22.8 km *to the nearest kilometer*. The distance in reality is about 23 km.

> 5.65 cm on map = 22.6 km in reality
>
> 5.688 cm on map = 22.752 km in reality
>
> 5.703 cm on map = 22.812 km in reality
>
> 5.718 cm on map = 22.872 km in reality
>
> 5.749 cm on map = 22.996 km in reality

Significant digits of a number are those digits whose value contributes to the precision of the number. Significant digits help us know how to round answers when calculating *measurements*, because measurements by their nature are never totally precise.

For example, all the individual digits of 12.593 m tell us something about its precision: it is precise to the thousandth of a meter. However, in the measurement 2,000 m, we cannot be sure if the number was originally measured as 1,9283.4 m and rounded to 2,000 m or measured as 2,400 m and rounded to 2,000 m. So in 2,000 m, only the 2 is a significant digit that tells us something about its precision.

All non-zero digits are always significant. With zeros, the situation is more complex. Here are the rules:

1. All non-zero digits are significant: 38.2 has three significant digits.

2. Zeros between other significant digits are also significant: 50,039 has five significant digits.

3. Non-decimal zeros at the end of a number are not significant: 6,400 has two significant digits.

4. Decimal zeros in front of the number are not significant: 0.0038 has two significant digits.

5. Decimal zeros at the end of a number *are* significant: 0.00380 has three significant digits.

In a calculation involving multiplication and/or division, the amount of significant digits in the answer should equal the amount of significant digits in the number that is the least precise (that has the smallest amount of significant digits).

For example, 2.3 cm has two significant digits and 11.9 cm has three. When we multiply them (to get an area), we get 27.37 cm^2, but we need to take the result to only *two* significant digits (because 2.3 cm had the least amount of significant digits, which was 2), so 27.37 cm^2 gets rounded to 27 cm^2.

In this lesson you will often multiply or divide a measurement result by a conversion factor. In this situation, **keep the same number of significant digits in your converted result as what you had in your measurement.** That is because the conversion factors are more exact and have more significant digits than your measurement result, so the measurement will automatically be the number with the least amount of significant digits.

You can use a calculator for all the problems in this lesson.

1. How many significant digits do these numbers have?

a. 24.5 km	**b.** 20.5 km	**c.** 24.50 km	**d.** 0.5 mi
e. 15,000 ft	**f.** 15,001 ft	**g.** 0.078 km	**h.** 0.0780 km
i. 5,002.90 kg	**j.** 340 lb	**k.** 340.9 lb	**l.** 0.005 lb

2. The two sides of a rectangular play area are measured to be 24.5 m and 13.8 m.

   a. Calculate its area and give it with a reasonable amount of significant digits.

   b. Let's say the dimensions of the play area were measured more accurately to be 24.56 m and 13.89 m. Calculate the area and give the result to a reasonable accuracy.

3. Calculate the following distances in reality. Consider how many significant digits your answer should have. Note: All digits in the scale ratios are significant. For example, the scale ratio of 1:50,000 is precise to all 5 digits. It's neither 1:49,999 nor 1:50,001, but exactly 1:50,000.

   a. 6.2 cm on a map with a scale of 1:50,000

   b. 12.5 cm on a map with a scale of 1:200,000

   c. 0.8 cm on a map with a scale of 1:15,000

4. A field measures 5.0 cm by 3.5 cm on a map with a scale of 1:8,000. Calculate its area in reality.

5. The distance from Mary's home to school is 3.0 inches on a map with a scale of 1:10,000.

   a. How long is this distance in reality? Give your answer in miles to two significant digits.

   b. Give your answer in yards to two significant digits.

6. A gas station is on a rectangular plot of land that measures 45.0 m by 31.2 m. What are these dimensions on a map with a scale of 1:500?

Review

1. Simplify the ratios and rates.

a. 164 km per 4 hours	**b.** $\dfrac{6\text{ g}}{1600\text{ ml}} =$	**c.** $52 : 156 =$ _____ : _____

2. A car traveled 348 miles in 6 hours. Fill in the table of equivalent rates.

Miles						348		
Hours	1	2	3	4	5	6	10	20

3. A mixture of salt and water contains 20 grams of salt and 1,200 grams of water.
   Write the ratio by weight of salt to water and simplify it.

4. Susan can jog 1 1/2 miles in 1/3 hour.
   Write a rate for her jogging speed and simplify it.

5. Solve the proportions. Round your answers to the nearest hundredth.

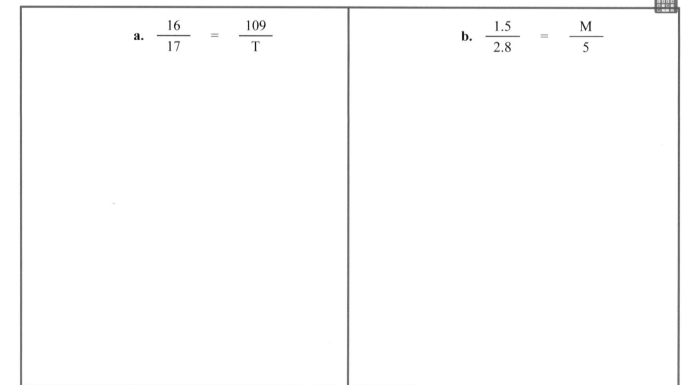

a. $\dfrac{16}{17} = \dfrac{109}{T}$

b. $\dfrac{1.5}{2.8} = \dfrac{M}{5}$

6. Write a proportion for the following problem and solve it.

12 kg of chicken feed costs $19.
How much would 5 kg cost?

_____ = _____

7. On the average, Gary makes a basket eight times out of every ten shots.
How many baskets can he expect to make when he practices 25 shots?

8. Write the unit rate as a complex fraction, and then simplify it.

a. Alex solved 2 1/2 pages of math problems in 1 1/4 hour.
b. Noah painted 2/3 of a room in 3/4 of an hour.

9. A car is traveling at a constant speed of 75 km per hour.

    a. Write an equation relating the distance (*d*) and time (*t*) and plot it in the grid below.

    b. What is the unit rate?

    c. Plot the point that matches the unit rate in this situation.

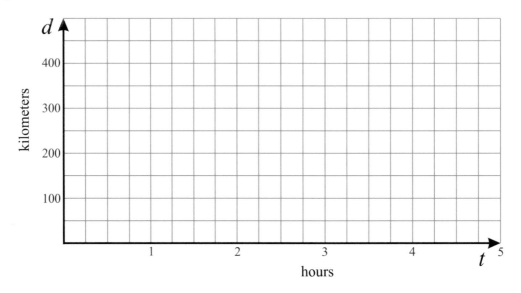

    d. What does the point (0, 0) mean in terms of this situation?

    e. How far can the car travel in 55 minutes, driving at the same speed?
       Also, plot the point for the time *t* = 55 min.

    f. How long will the car take to travel 160 km? Give your answer in hours and minutes.
       Also, plot the point that matches the distance *d* = 160 km.

10. Using a pre-paid internet service you get a certain amount of bandwidth to use for the amount you pay. The table shows the prices for certain amounts of bandwidth.

Bandwidth	1G	2G	5G	10G	15G	20G	25G
Price	$10	$16	$23	$30	$37	$43	$50

a. Are these two quantities in proportion?

Explain how you can tell that.

b. If so, write an equation relating the two and state the constant of proportionality.

11. In the year 2008 it was estimated that it cost $9,369 a year to drive a medium-sized car (a sedan) for 15,000 miles (a typical amount of use). Based on those same assumptions, how much would it cost, to the nearest dollar, to drive that car for 5 months?

12. The figures are similar. Find the length of the side labeled with x.

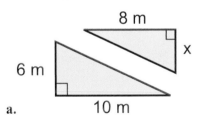

a.

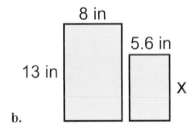

b.

13. A house plan has a scale of 1 in : 6 ft. In the plan, one room measures 2 in × 2 ¾ in. What are the true dimensions of the room?

14. A freight truck fully loaded with cargo gets six miles to a gallon of diesel.

   a. What is the unit rate in this situation?

   b. Write an equation relating the mileage (M) to the amount of diesel fuel (*f*) in gallons.

   c. Plot your equation. Choose an appropriate scaling for the two axes.

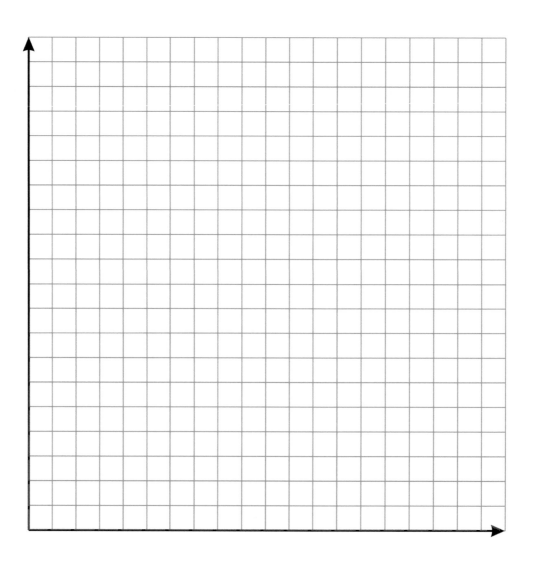

Ratios and Proportions Answer Key

Ratios and Rates, p. 13

1. The ratio of triangles to diamonds is 8:10 = 4:5.
   In this picture, there are 4 triangles to every 5 diamonds.

2. a.    b. The ratio of circles to pentagons is 2:7.

3. a.    b. The ratio of diamonds to triangles is 9:15 or 3:5.

4.

a. 5 to 45 = 1 to 9	b. 3:20 = 9:60	c. 280:420 = 2:3	d. $\dfrac{5}{13} = \dfrac{\boxed{25}}{65}$

5.

a. 5 kg and 800 g	b. 600 cm and 2.4 m
$\dfrac{5\text{ kg}}{800\text{ g}} = \dfrac{5{,}000\text{ g}}{800\text{ g}} = \dfrac{5000}{800} = \dfrac{25}{4}$	$\dfrac{600\text{ cm}}{2.4\text{ m}} = \dfrac{600\text{ cm}}{240\text{ cm}} = \dfrac{600}{240} = \dfrac{5}{2}$
c. 1 gallon and 3 quarts	d. 3 ft 4 in and 1 ft 4 in
$\dfrac{1\text{ gal}}{3\text{ qt}} = \dfrac{4\text{ qt}}{3\text{ qt}} = \dfrac{4}{3}$	$\dfrac{3\text{ ft 4 in}}{1\text{ ft 4 in}} = \dfrac{40\text{ in}}{16\text{ in}} = \dfrac{40}{16} = \dfrac{5}{2}$

6.

a. 5.6 km and 3.2 km	b. 0.02 m and 0.5 m
$\dfrac{5.6\text{ km}}{3.2\text{ km}} = \dfrac{5600\text{ m}}{3200\text{ m}} = \dfrac{56}{32} = \dfrac{7}{4} = 7{:}4$	$\dfrac{0.02\text{ m}}{0.5\text{ m}} = \dfrac{2\text{ cm}}{50\text{ cm}} = \dfrac{2}{50} = \dfrac{1}{25} = 1{:}25$
c. 1.25 m and 0.5 m	d. 1/2 L and 7 1/2 L
$\dfrac{1.25\text{ m}}{0.5\text{ m}} = \dfrac{125\text{ m}}{50\text{ m}} = \dfrac{125}{50} = \dfrac{5}{2} = 5{:}2$	$\dfrac{1/2\text{ L}}{7\ 1/2\text{ L}} = \dfrac{1\text{ L}}{15\text{ L}} = \dfrac{1}{15} = 1{:}15$
e. 1/2 mi and 3 1/2 mi	f. 2/3 km and 1 km
$\dfrac{1/2\text{ mi}}{3\ 1/2\text{ mi}} = \dfrac{1\text{ mi}}{7\text{ mi}} = \dfrac{1}{7} = 1{:}7$	$\dfrac{2/3\text{ km}}{1\text{ km}} = \dfrac{2\text{ km}}{3\text{ km}} = \dfrac{2}{3} = 2{:}3$

7. a. Jeff swims at a constant speed of 1,200 ft : 15 minutes, 1,200 ft per 15 min, or 1,200 ft / 15 min.
      The simplified rate is 80 ft : 1 min or 80 ft per minute.

   b. The car can travel 54 mi : 3 gal, 54 mi per 3 gal, or 54 mi / 3 gal.
      The simplified rate is 18 mi : 1 gal or 18 miles per gallon.

8.

a. $\dfrac{1/2\text{ cm}}{30\text{ min}} = \dfrac{1\text{ cm}}{1\text{ hr}} = \dfrac{1/4\text{ cm}}{15\text{ min}}$	b. $\dfrac{\$88.40}{8\text{ hr}} = \dfrac{\$22.10}{2\text{ hr}} = \dfrac{\$110.50}{10\text{ hr}}$

Ratios and Rates, cont.

9.

a.	$\dfrac{280 \text{ km}}{7 \text{ hr}} = \dfrac{140 \text{ km}}{3.5 \text{ hr}} = \dfrac{40 \text{ km}}{1 \text{ hr}}$	b.	$\dfrac{2.5 \text{ in}}{1.5 \text{ min}} = \dfrac{25 \text{ in}}{15 \text{ min}} = \dfrac{5 \text{ in}}{3 \text{ min}}$

10.

Distance (km)	12	36	48	60	72	108	120
Time (min)	10	30	40	50	60	90	100

11. Always look for the easy way to solve these rate problems. Here, you are given a rate of 8 pairs of socks for $20. Half of 8 is 4, and half of $20 is $10. Figuring half again gives 2 for $5. Half of that is 1 for $2.50. With those rates, you can easily calculate the rest. Since 6 = 4 + 2, the rate for 6 is $10 + $5 = $15. Since 7 = 6 + 1, the rate for 7 is $15 + $2.50 = $17.50. Since 9 = 8 + 1, the rate for 9 is $20 + $2.50 = $22.50. And since 10 = 8 + 2, the rate for 10 is just $20 + $5 = $25.

Cost ($)	$2.50	$5	$10	$15	$17.50	$20	$22.50	$25
Pairs of socks	1	2	4	6	7	8	9	10

Solving Problems Using Equivalent Rates, p. 16

1.

a. Distance	15 km	5 km	1.25 km	3.75 km
Time	3 hr	1 hr	15 min	45 min

b. Pay	$6	$2	$8	$14
Time	45 min	15 min	1 hr	1 hr 45 min

2. a. $\dfrac{3 \text{ in}}{8 \text{ ft}} = \dfrac{0.75 \text{ in}}{2 \text{ ft}} = \dfrac{4.5 \text{ in}}{12 \text{ ft}} = \dfrac{7.5 \text{ in}}{20 \text{ ft}}$ b. $\dfrac{115 \text{ words}}{2 \text{ min}} = \dfrac{57.5}{1 \text{ min}} = \dfrac{172.5}{3 \text{ min}}$

3. Jake will take 63 minutes to ride 36 miles. $\dfrac{8 \text{ miles}}{14 \text{ minutes}} = \dfrac{4 \text{ miles}}{7 \text{ minutes}} = \dfrac{36 \text{ miles}}{63 \text{ minutes}}$

4. Larry will have to work 46 2/3 hours or 46 hours 40 minutes to earn $600.
You can first divide both terms of the rate $90 : 7 hr by three to get $30 : 7/3 hr. Double those to get $60 : (4 ⅔ hr). Then multiply the terms of that rate by ten to get $600 : (40 hr + 20/3 hr) = $600 : 46 ⅔ hr.

Earnings	$30	$60	$90	$600
Work Hours	7/3 hr	4 2/3 hr	7 hr	46 ⅔ hr

5. The rate 45 mi : 2 gal is equal to 15 mi : 2/3 gal. Now, multiply the terms of that rate by four to get 60 mi : 8/3 gal. So the car would use 2 ⅔ gallons of gasoline for a 60-mile trip.

6. a. $\dfrac{80 \text{ miles}}{60 \text{ min}} = \dfrac{80/3 \text{ miles}}{20 \text{ min}} = \dfrac{560/3 \text{ miles}}{140 \text{ min}} = \dfrac{186 \text{ ⅔ miles}}{140 \text{ min}}$. The train can go 186 ⅔ miles in 140 minutes.

  b. No, because 50 miles in 40 minutes is equal to 25 miles in 20 minutes, we get 75 miles in 60 minutes, which is less than the 80 miles in 60 minutes above.

7. No, 30 pencils for $4.50 is equal to the rate of 50 pencils for $7.50.

Price	$4.50	$1.50	$7.50
Pencils	30	10	50

8. a. 640:1,000 = 16:25

  b. You can multiply both terms in the rate 16:25 by five to get 80:125. So we would expect 80 people in a group of 125 people would like blue the best.

Solving Proportions: Cross Multiplying, p. 18

1.

a. $\dfrac{15}{32} = \dfrac{67}{M}$ First cross-multiply.

$15M = 32 \cdot 67$ This is the equation you get after cross-multiplying.

$15M = 2{,}144$ In this step, calculate what is on the right side.

$\dfrac{15M}{15} = \dfrac{2144}{15}$ In this step, divide both sides of the equation by _15_.

$M \approx 142.9$ This is the final answer.

b. $\dfrac{7}{146} = \dfrac{38}{S}$ First cross-multiply.

$7S = 146 \cdot 38$ This is the equation you get after cross-multiplying.

$7S = 5{,}548$ In this step, calculate what is on the right side.

$\dfrac{7S}{7} = \dfrac{5548}{7}$ In this step, divide both sides of the equation by _7_.

$S \approx 792.6$ This is the final answer.

2.

a. $\dfrac{1.2}{4.5} = \dfrac{G}{7.0}$ First cross-multiply.

$1.2 \cdot 7 = 4.5G$
$4.5G = 1.2 \cdot 7$ (You may want to flip the sides so the variable is on the left.)

$4.5G = 8.4$ Calculate what is on the right side.

$\dfrac{4.5G}{4.5} = \dfrac{8.4}{4.5}$ In this step, divide both sides of the equation by _4.5_.

$G \approx 1.9$ This is the final answer.

b. $\dfrac{4.3}{C} = \dfrac{10}{17}$ First cross-multiply.

$4.3 \cdot 17 = 10C$
$10C = 4.3 \cdot 17$ (You may want to flip the sides so the variable is on the left.)

$10C = 73.1$ Calculate what is on the right side.

$\dfrac{10C}{10} = \dfrac{73.1}{10}$ In this step, divide both sides of the equation by _10_.

$C \approx 7.3$ This is the final answer.

3.

a. $\dfrac{T}{25} = \dfrac{15}{3}$

$3T = 25 \cdot 15$

$3T = 375$

$\dfrac{3T}{3} = \dfrac{375}{3}$

$T = 125$

b. $\dfrac{17}{214} = \dfrac{2}{M}$

$17M = 214 \cdot 2$

$17M = 428$

$\dfrac{17M}{17} = \dfrac{428}{17}$

$M \approx 25.18$

4.

a. Let P be the amount of paint she needs to paint 240 m².	b. Let M be the distance the car goes on 7.9 gallons of gasoline.

a.

$$\frac{P}{240 \text{ m}^2} = \frac{85 \text{ L}}{700 \text{ m}^2}$$

$$700 \text{ m}^2 \cdot P = 85 \text{ L} \cdot 240 \text{ m}^2$$

$$700 \text{ m}^2 \cdot P = 20{,}400 \text{ L} \cdot \text{m}^2$$

$$\frac{700 \text{ m}^2 \cdot P}{700 \text{ m}^2} = \frac{20{,}400 \text{ L} \cdot \text{m}^2}{700 \text{ m}^2}$$

$$P \approx 29 \text{ L}$$

So May needs about 29 liters of paint.

Since 700 m² is a little less than three times 240 m², and 85 liters is a little less than three times 29 liters, the answer is reasonable.

b.

$$\frac{M}{7.9 \text{ gal}} = \frac{56.0 \text{ mi}}{3.2 \text{ gal}}$$

$$3.2 \text{ gal} \cdot M = 7.9 \text{ gal} \cdot 56.0 \text{ mi}$$

$$3.2 \text{ gal} \cdot M = 442.4 \text{ gal} \cdot \text{mi}$$

$$\frac{3.2 \text{ gal} \cdot M}{3.2 \text{ gal}} = \frac{442.4 \text{ gal} \cdot \text{mi}}{3.2 \text{ gal}}$$

$$M \approx 138 \text{ mi}$$

The car can travel about 138 miles on 7.9 gallons.

Since 7.9 gal is more than twice but less than three times 3.2 gal, and 138 miles is more than twice but less than three times 56.0 miles, the answer is reasonable.

5. Either way gives the correct answer of $x \approx 32$ gallons. Notice that Noah's data is inverted to Jack's proportion.

A car travels 154 miles on 5.5 gallons of gasoline. How many gallons would it need to travel 900 miles?

Jack's proportion:

$$\frac{154 \text{ mi}}{5.5 \text{ gal}} = \frac{900 \text{ mi}}{x}$$

$$154 \text{ mi} \cdot x = 900 \text{ mi} \cdot 5.5 \text{ gal}$$

$$154 \text{ mi} \cdot x = 4{,}950 \text{ mi} \cdot \text{gal}$$

$$\frac{154 \text{ mi} \cdot x}{154 \text{ mi}} = \frac{4{,}950 \text{ mi} \cdot \text{gal}}{154 \text{ mi}}$$

$$P \approx 32 \text{ gal}$$

Noah's proportion:

$$\frac{5.5 \text{ gal}}{154 \text{ mi}} = \frac{x}{900 \text{ mi}}$$

$$5.5 \text{ gal} \cdot 900 \text{ mi} = 154 \text{ mi} \cdot x$$

$$154 \text{ mi} \cdot x = 4{,}950 \text{ mi} \cdot \text{gal}$$

$$\frac{154 \text{ mi} \cdot x}{154 \text{ mi}} = \frac{4{,}950 \text{ mi} \cdot \text{gal}}{154 \text{ mi}}$$

$$P \approx 32 \text{ gal}$$

6. (b) is set up correctly.

b. $$\frac{23 \text{ m}}{3 \text{ s}} = \frac{x}{7 \text{ s}}$$

$$23 \text{ m} \cdot 7 \text{ s} = 3 \text{ s} \cdot x$$

$$3 \text{ s} \cdot x = 161 \text{ m} \cdot \text{s}$$

$$\frac{3 \text{ s} \cdot x}{3 \text{ s}} = \frac{161 \text{ m} \cdot \text{s}}{3 \text{ s}}$$

$$x = 53 \; 2/3 \text{ m} \approx 54 \text{ m}$$

7. Seventeen kilograms of dog food costs $43.35. Since 20 kg of dog food costs $51, we would expect 17 kg to cost less than $51 but reasonably close to it, around $40 or more. So our answer is reasonable.

Jane's proportion is not set up correctly because on the left side kilograms are on the top (in the numerator), but on the right side they are on the bottom (in the denominator). She would end up with x in units of kg^2 / $\$$, which is wrong.

Jane's proportion:	The corrected proportion:
$$\frac{20 \text{ kg}}{\$51} = \frac{x}{17 \text{ kg}}$$	$$\frac{\$51}{20 \text{ kg}} = \frac{x}{17 \text{ kg}}$$ $$\$51 \cdot 17 \text{ kg} = 20 \text{ kg} \cdot x$$ $$20 \text{ kg} \cdot x = \$867 \cdot 1 \text{ kg}$$ $$\frac{20 \text{ kg} \cdot x}{20 \text{ kg}} = \frac{\$867 \cdot 1 \text{ kg}}{20 \text{ kg}}$$ $$x = \$43.35$$

8.

a. You need 3.6 pounds of fertilizer. The lawn is 30 ft × 24 ft = 720 ft², so the proportion is:	b. Six liters of paint would cover about 49 m².
$$\frac{5 \text{ lb}}{1000 \text{ ft}^2} = \frac{x}{720 \text{ ft}^2}$$ $$5 \text{ lb} \cdot 720 \text{ ft}^2 = 1000 \text{ ft}^2 \cdot x$$ $$1000 \text{ ft}^2 \cdot x = 3600 \text{ lb} \cdot \text{ft}^2$$ $$\frac{1000 \text{ ft}^2 \cdot x}{1000 \text{ ft}^2} = \frac{3600 \text{ lb} \cdot \text{ft}^2}{1000 \text{ ft}^2}$$ $$x = 3.6 \text{ lb}$$	$$\frac{700 \text{ m}^2}{85 \text{ L}} = \frac{x}{6 \text{ L}}$$ $$700 \text{ m}^2 \cdot 6 \text{ L} = 85 \text{ L} \cdot x$$ $$85 \text{ L} \cdot x = 4{,}200 \text{ m}^2 \cdot \text{L}$$ $$\frac{85 \text{ L} \cdot x}{85 \text{ L}} = \frac{4200 \text{ m}^2 \cdot \text{L}}{85 \text{ L}}$$ $$x \approx 49.4 \text{ m}^2$$

Why Cross-Multiplying Works, p. 24

1.

a. $\dfrac{8}{1.15} = \dfrac{37}{K}$	b. $\dfrac{8}{1.15} = \dfrac{37}{K}$ Multiply both sides by K.
$8K = 37 \cdot 1.15$	$\dfrac{8}{1.15} \cdot K = \dfrac{37}{K} \cdot K$ K cancels on the right side.
$8K = 42.55$	$\dfrac{8}{1.15} \cdot K = 37$ Multiply both sides by 1.15.
$\dfrac{8K}{8} = \dfrac{42.55}{8}$	$\dfrac{8}{1.15} \cdot K \cdot 1.15 = 37 \cdot 1.15$ Simplify.
$K = 5.31875$	$8K = 42.55$ Divide both sides by 8.
	$\dfrac{8K}{8} = \dfrac{42.55}{8}$
	$K = 5.31875$

Unit Rates, p. 25

1. a. $125 for 5 packages = <u>$25 per package</u>

   b. $6 for 30 envelopes = <u>$0.20 per envelope</u>

   c. $1.37 for 1/2 hour = <u>$2.74 per hour</u>

   d. 2 1/2 inches per 4 minutes = 5/2 inches per 4 minutes = <u>5/8 inch per minute</u>

   e. 24 m^2 per 3/4 gallon = 24 ÷ (3/4) = 24 · (4/3) = <u>32 m^2 per gallon</u>

2. $\dfrac{\frac{1}{2}}{\frac{1}{4}}$ miles per hour = (1/2) ÷ (1/4) mph = (1/2) · 4 mph = 2 mph

3.

a.	$\dfrac{5\ 1/2\ \text{yd}}{3\ \text{skirts}} = \dfrac{11/2\ \text{yd}}{3\ \text{skirts}} = (11/2) \cdot (1/3)$ yd/skirt $= 11/6$ yd/skirt $= 1\ 5/6$ yd/skirt
	She uses 1 5/6 yards of material per skirt.
b.	$\dfrac{2\ 3/4\ \text{servings}}{30\ \text{g}} = \dfrac{11/4\ \text{servings}}{30\ \text{g}} = 11/4 \cdot (1/30)$ servings per gram $= 11/120$ servings per gram
	One gram of powder gives you 11/120 serving of vegetables.
c.	$\dfrac{11/4\ \text{mi}}{5/6\ \text{h}} = 11/4 \cdot (6/5)$ mph $= 66/20$ mph $= 33/10$ mph $= 3\ 3/10$ mph
	Marsha walked at a speed of 3 3/10 or 3.3 miles per hour.

3.

d.	$\dfrac{1\ 1/2\ \text{vases}}{2\ 1/2\ \text{hours}} = \dfrac{3/2\ \text{vases}}{5/2\ \text{h}} = 3/2 \cdot (2/5)$ vases per hour $= 3/5$ vase per hour - or - $\dfrac{2\ 1/2\ \text{hours}}{1\ 1/2\ \text{vases}} = \dfrac{5/2\ \text{h}}{3/2\ \text{vases}} = 5/2 \cdot (2/3)$ hours per vase $= 5/3$ hours per vase Linda makes 3/5 of a vase per hour and takes 5/3 hour per vase.
e.	$\dfrac{5,400\ \text{people}}{3/8\ \text{km}^2} = 5,400 \cdot (8/3)$ people per km^2 $= 14,400$ people per km^2 There are 14,400 people in one square kilometer.
f.	$\dfrac{\$8.70}{5/8\ \text{lb}} = \$8.70 \cdot (8/5)$ per lb $= \$13.92$ per lb The nuts cost $13.92 per pound.
g.	$\dfrac{3/8\ \text{game}}{7/12\ \text{h}} = 3/8 \cdot (12/7)$ games per hour $= 36/56$ games per hour $= 9/14$ games per hour - or - $\dfrac{7/12\ \text{h}}{3/8\ \text{game}} = 7/12 \cdot (8/3)$ hours per game $= 56/36$ hours per game $= 14/9$ hours per game $= 1\ 5/9$ hours per game He plays 9/14 of a game per hour, and he takes 1 5/9 hours or 1 hour 33 minutes to finish one game.

4. a. $d = 50t$

b.
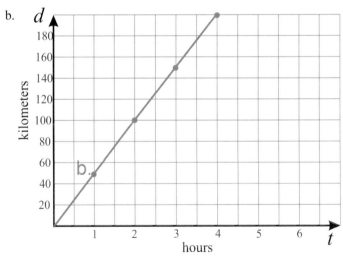

c. It means that at the time of 3 hours the truck has traveled 150 kilometers.

5. a. d = (1/3)t

   b. The unit rate is (1/3 m)/(1 s), which can also be written as 1/3 m/s.

   c. e. f.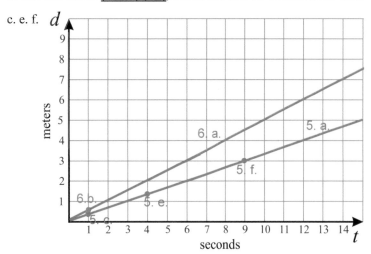

   d. The point (0, 0) means that at the time 0 seconds, the duck has traveled 0 meters. That point is the starting point.

6. a. - b. See the image above.
   c. The line for the adult ducks rises quicker or is steeper than the line for the babies.
   d. At t = 5s, the adult ducks have walked 5/2 meters and the babies have walked 5/3 meters.
      The difference is 5/2 m − 5/3 m = (15/6 − 10/6) = 5/6 m.
   e. It will take 15 s for the baby ducks to walk 5 meters and 10 s for the adult ducks so the babies
      will take 5 s longer than the adults to walk 5 meters.

7. a. To calculate the rate of doctors per 10,000 people, we need to find how many groups of 10,000 people there are
      in the whole population: 7,365,000 ÷ 10,000 = 736.5. So if we put 736.5 (groups of 10,000 people) as the second
      term of the rate and convert it to a unit rate, we will find how many doctors there are per one group of 10,000
      people. There are 27,700 / 736.5 ≈ 37.6 doctors per 10,000 people.

   b. Since 350,000 people is 35 groups of 10,000, and since there is an average of 12.1 doctors per 10,000 people,
      we would expect to find 35 · 12.1 = 423.5 physicians in an area with 350,000 residents.

   c. The rate is 1.2 physicians per 1,000 people.

8. a. Jane: 38 beats in 15 seconds = 152 beats in 60 seconds = 38/15 beats per second (about 2.53 beats per second)
      Stacy: 52 beats in 20 seconds = 156 times in 60 seconds = 13/5 beats per second (2.6 beats per second)

      So Stacy's heart rate is faster by 4 heartbeats per minute.

   b. Let H be the number of heartbeats and t time in seconds. For Jane, the equation is H = (38/15)t
      and the unit rate is 38/15 beats per second.

   c. For Stacy, the equation is H = (13/5)t and the unit rate is 13/5 beats per second.

1. a. $y = 3x$

   Yes, the variables are in direct variation.

y	−9	−6	−3	0	3	6	9	12
x	−3	−2	−1	0	1	2	3	4

b. $y = x + 2$

   No, the variables are not in direct variation.

y	−1	0	1	2	3	4	5	6
x	−3	−2	−1	0	1	2	3	4

c. $y = (1/2)x − 1$

   No, the variables are not in direct variation.

y	−2.5	−2	−1.5	−1	−0.5	0	0.5	1
x	−3	−2	−1	0	1	2	3	4

d. $y = −x$

   Yes, the variables are in direct variation.

y	3	2	1	0	−1	−2	−3	−4
x	−3	−2	−1	0	1	2	3	4

e. $C = 2.4n$

   Yes, the variables are in direct variation.

C	0	2.4	4.8	7.2	9.6	12	14.4	16.8
n	0	1	2	3	4	5	6	7

f. $h = 1/k$

   No, the variables are not in direct variation.

h	1	1/2	1/3	1/4	1/5	1/6	1/7	1/8
k	1	2	3	4	5	6	7	8

2. a. $y = x + 1$ No.

x	−3	−2	−1	0	1	2	3	4
y	−2	−1	0	1	2	3	4	5

b. $y = 2x$ Yes.

x	−3	−2	−1	0	1	2	3	4
y	−6	−4	−2	0	2	4	6	8

c. $y = 2x − 1$ No.

x	−3	−2	−1	0	1	2	3	4
y	−7	−5	−3	−1	1	3	5	7

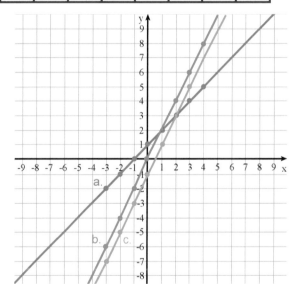

d. $y = (1/2)x$ Yes.

x	−3	−2	−1	0	1	2	3	4
y	−1.5	−1	−0.5	0	0.5	1	1.5	2

e. $y = −2x$ Yes.

x	−3	−2	−1	0	1	2	3	4
y	6	4	2	0	−2	−4	−6	−8

f. $y = −2x + 1$ No.

x	−3	−2	−1	0	1	2	3	4
y	7	5	3	1	−1	−3	−5	−7

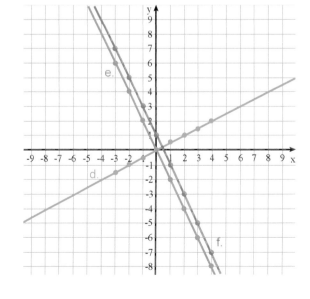

Proportional Relationships, cont.

3. Please check the student's answers.
   If the variables are in direct variation, the equation is of the form $y = mx$, and if they are not in direct variation, the equation is of the form $y = mx + c$.
   If the variables are in direct variation, the plot shows a line going through the origin, and otherwise not.

4. a. See the graph at the right.
   b. The unit rate is 3:1.
   c. See the graph at the right.

5. a. See the graph at the right.
   b. The unit rate is 0.5:1.
   c. See the graph at the right.

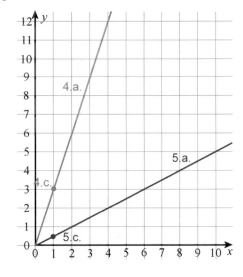

6. a.

x	−3	−2	−1	0	1	2	3	4
y	−5	−4	−3	−2	−1	0	1	2

In proportion or not? No.

b.

x	−3	−2	−1	0	1	2	3	4
y	−12	−8	−4	0	4	8	12	16

In proportion or not? Yes
Unit rate: 4:1
Equation: $y = 4x$

c.

x	−3	−2	−1	0	1	2	3	4
y	−1	−2/3	−1/3	0	1/3	2/3	1	4/3

In proportion or not? Yes
Unit rate: 1/3:1
Equation: $y = (1/3)x$

d.

x	−3	−2	−1	0	1	2	3	4
y	2	5/3	4/3	1	2/3	1/3	0	−1/3

In proportion or not? No.

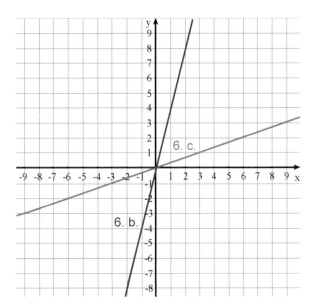

7. a. (1) Unit Rate: 15 m per second (or (15 m)/(1 s) or 15 m/s).
   (2) See the graph on the right.
   (3) Equation: $d = 15t$

   b. (1) Unit rate: 10/3 m per second = 3 1/3 m per second (or 10/3 m/s).
   (2) See the graph on the right.
   (3) Equation: $d = (10/3)t$

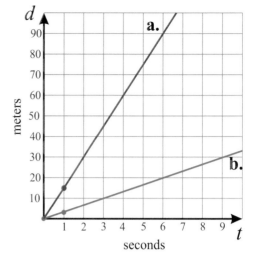

8. a. Equation: $V = (1/10)A$

   b. Equation: $V = (1/15)A$

   c. These could represent two different paints. The graphs would show how much paint is needed to cover a certain area.

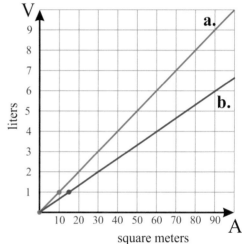

9. a. $d = 45t$
   b. See the graph on the right.
   c. The point is (1, 45). See the graph on the right.

10. a. The quantities are NOT in direct proportion. The relation doesn't include the point (0,0).

11. a. These two quantities are in proportion.
    b. Since three workers can plant 60 trees on average, one worker can plant 20. This gives us the unit rate, 20 trees per worker (in one day). Let w be the number workers and T be the number of trees they can plant (in one day). The equation is $\underline{T = 20w}$. The constant of proportionality is 20.

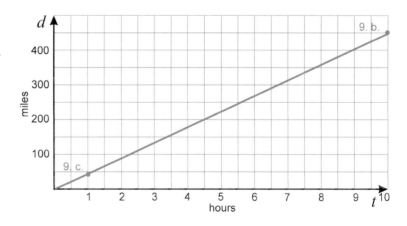

12. a. The quantities are NOT in direct proportion. When one goes up, the other comes down.
    b. Interestingly, this data is in what is called an *inverse proportion* between the time (t) and number of workers (N). It takes 10 hours of work to paint the house. When there is only one worker, he takes the full 10 hours to finish. When there are two workers, each takes 5 hours, and together they finish the ten-hour job. When there are three workers, each does 3.3 hours of the ten-hour job, and so on. The equation is $t = 10/N$, which is NOT in the form $y = mx$.

Proportional Relationships, cont.

13.a. The two quantities, the cost C and the calling time t, are not in proportion because, for one reason, the point $(0, 0)$ is not included in the graph. Also, the graph is not a straight line through origin but consists of two different lines: one horizontal and one rising.

    b. The coordinates of the marked point are $(7, 35)$.

    c. That means it costs $35 for seven hours of calling time.

Graphing Proportional Relationships–More Practice, p. 36

1. a. The unit rate is $400/7 \approx 57$ ml per month.

    b. $S = (400/7)t$.
    (We'd normally write the equation using as $S = (400/7)t$ because $400/7$ is exact. Writing $S \approx 57t$, $S \approx 57.1t$, $S \approx 57.14t$, or $S \approx 57.143t$ is only approximate.)

    c. - e. See the graph on the right.

    f. We can use the equation $S = (400/7)t$ to find out how long it takes her to use 250 ml of shampoo. Simply set S to be 250 and solve the equation:

$$250 = (400/7)t$$

$$(400/7)t = 250$$

$$400t = 250 \cdot 7$$

$$t = \frac{1,750}{400}$$

$$t = 4.375$$

It would take her about 4 months 11 days to use 250 ml of shampoo.

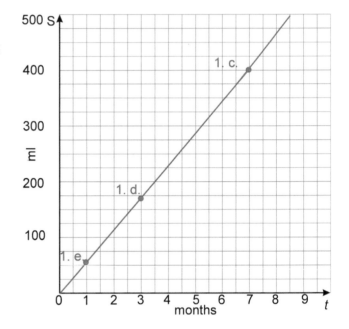

2. a. Earning $100 in 8 hours means earning $12.50 in one hour, so our equation is $\underline{P = \$12.50t}$

    b. See the graph on the right. The point is $(20, 250)$.

    c. See the graph on the right. The point for the unit rate is $(1, 12.5)$.

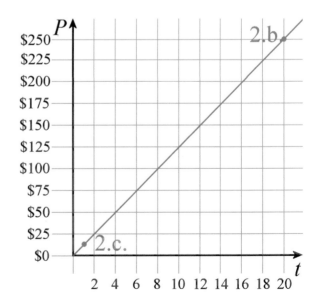

Graphing Proportional Relationships–More Practice, cont.

3. a. $m = 28g$

   b. See the graph at the right.

   c. Point (3.6, 100). See the graph at the right.

   d. Point (10, 280). See the graph at the right.

   e. Point (1, 28). See the graph at the right.

   f. It means when the car has gone zero miles it has used zero gallons of gas, or that with zero gallons of gas, the car can travel zero miles.

   g. The car can travel 2,800 miles on 100 gallons of gasoline.

   h. The car can travel 933 1/3 miles on \$100 of gasoline. At \$3 a gallon, \$100 buys 33 1/3 gallons, so the mileage is m = (28 miles/gallon) · (33 1/3 gallons) = 933 1/3 miles.

   i. To travel 700 miles, the car uses 700/28 = 25 gallons of gasoline, which would cost <u>\$75</u>.

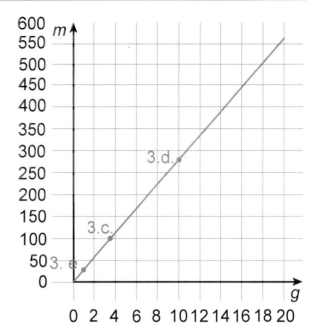

More on Proportions, p. 38

1. Five books would weigh 7.6 pounds. Some example solutions:

Solution with a proportion:

$$\frac{27.4 \text{ lb}}{18 \text{ books}} = \frac{x}{5 \text{ books}}$$

$$27.4 \cdot 5 = 18x$$

$$18x = 137$$

$$x = \frac{137}{18}$$

$$x \approx 7.6$$

Solution with a unit rate:

If 18 books weigh 27.4 lb, then one book weighs $27.4 \div 18 = 1.5\overline{2}$ lb.

And five books weigh:
$5 \cdot 1.5\overline{2} = 7.6\overline{1}$ lb $\approx$ 7.6 lb.

Solution in a diagram form:

÷ 18	18 books	27.4 lb	÷ 18
	1 book	$1.5\overline{2}$ lb	
· 5	5 books	$7.6\overline{1}$ lb	· 5

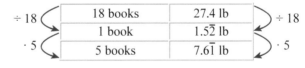

2. Solutions may vary. Check the student's work. Here are some examples:

Solution with a proportion:

$$\frac{\$45}{8 \text{ m}} = \frac{x}{70 \text{ m}}$$

$$\$45 \cdot 70 \text{ m} = x \cdot 8 \text{ m}$$

$$8x \text{ m} = 3160 \ \$ \cdot \text{m}$$

$$\frac{8x \text{ m}}{8 \text{ m}} = \frac{3160 \ \$ \cdot \text{m}}{8 \text{ m}}$$

$$x = \$393.75$$

Solution by logical reasoning:

The perimeter of the plot is what gets fenced, and the perimeter is 15 m + 20 m + 15 m + 20 m = 70 m. Since 70 ÷ 8 = 8.75, you'll need to purchase eight complete rolls and part of a ninth. The eight complete rolls give you 8 · 8 m = 64 m and cost 8 · \$45 = \$360. You still need 70 m − 64 m = 6 m. That will cost (6 m / 8 m) · \$45 = \$33.75. So the total cost is \$360 + \$33.75 = <u>\$393.75</u>.

Solution with a unit rate:

The price per meter is \$45 ÷ 8 = \$5.625. The total price is then 70 m · \$5.625/m = <u>\$393.75</u>.

More on Proportions, cont.

3. Solutions may vary. Check the student's work. Here are some examples:

(1) Solution using a proportion:	(2) Solution using a unit rate:
$$\frac{s}{60 \text{ min}} = \frac{15 \text{ km}}{37 \text{ min}}$$ $$37 \text{ min} \cdot s = 60 \text{ min} \cdot 15 \text{ km}$$ $$37 \text{ min} \cdot s = 900 \text{ min} \cdot \text{km}$$ $$s = \frac{900 \text{ min} \cdot \text{km}}{37 \text{ min}}$$ $$s \approx 24.3 \text{ km}$$	The speed of 15 km in 37 minutes means (15/37) km per minute. In one hour, the boy can then ride $60 \cdot (15/37)$ km $= 24.\overline{324}$ km $\approx$ <u>24.3 km</u>.

4. The pasta sauce that costs $3.95 for 450 g costs more per gram.

(1) Solution using a proportion:	(2) Solution using unit rates:
If the rates are equal, they are in proportion, and cross-multiplying will produce a true equation. $$\frac{\$3.95}{450 \text{ g}} \overset{?}{=} \frac{\$4.55}{560 \text{ g}}$$ $$\$3.95 \cdot 560 \text{ g} \overset{?}{=} \$4.55 \cdot 450 \text{ g}$$ $$2{,}212 \cdot \text{g} \neq 2{,}548 \cdot \text{g}$$ We didn't get a true equation, so the rates are not equal.	Let's figure in kilograms instead of grams, so our unit rates are bigger numbers. The first unit rate is $3.95/0.450 kg = $8.78/kg. The second is $4.55/0.560 kg = $8.12/kg. The unit rates are different, so the original rates are, too. Other things being equal, the sauce in the 560g container is a better value for money. You can also calculate the unit rate in grams. However, you will get such small numbers that unit rates in kilograms are preferable. The first unit rate is $3.95/450 g = 0.008$\overline{7}$ g. The second is $4.55/560 g = 0.008125 g.

5. Yes, the rates are equal.

(1) Solution using a proportion:	(2) Solution using the unit rates:
If the rates are equal, they are in proportion, and cross-multiplying will produce a true equation. $$\frac{50 \text{ mi}}{2.2 \text{ gal}} \overset{?}{=} \frac{125 \text{ mi}}{5.5 \text{ gal}}$$ $$50 \text{ mi} \cdot 5.5 \text{ gal} \overset{?}{=} 125 \text{ mi} \cdot 2.2 \text{ gal}$$ $$275 \text{ mi} \cdot \text{gal} = 275 \text{ mi} \cdot \text{gal}$$ We got a true equation, so the rates are equal.	The first unit rate is 50 mi/2.2 gal = 22.$\overline{72}$ gal. The second is 125 mi/5.5 gal = 22.$\overline{72}$ gal. The unit rates are equal, so the original rates are equal.
	(3) Solution using logical reasoning:
	125 mi is 2 ½ times 50 mi, and 5.5 is 2 ½ times 2.2. So the ratios are the same.

More on Proportions, cont.

6. Answers will vary. Check the student's work. Possibly something like:

a. Word problem:	**b.** Word problem:
"Jared measured his typing speed for a certain passage that had 60 words. It took him 97 seconds. Typing at the same speed, how long would Jared take to type a novella of 10,000 words?"	"Lola can buy cardamom spice for $5.59 for 80 grams. At that same rate, how much cardamom can she get for $100.00?"
a. Proportion:	**b.** Proportion:

a. Proportion:

$$\frac{60 \text{ words}}{97 \text{ sec}} = \frac{10{,}000 \text{ words}}{x}$$

$$60 \text{ words} \cdot x = 970{,}000 \text{ words} \cdot \text{sec}$$

$$x = \frac{970{,}000 \text{ words} \cdot \text{sec}}{60 \text{ words}}$$

$$x = 16{,}166.\overline{6} \text{ sec}$$

It would take him about 16,167 seconds, which is about 4 hours 29 minutes. Since 97 is a little over 1 ½ times 60, and 16,167 is a little over 1 ½ times 10,000, that answer is reasonable.

b. Proportion:

$$\frac{\$5.59}{80 \text{ g}} = \frac{\$100}{x}$$

$$\$5.59 \cdot x = 80 \text{ g} \cdot \$100$$

$$x = \frac{80 \text{ g} \cdot \$100}{\$5.59}$$

$$x \approx 1{,}430 \text{ g}$$

She can get about 1,430 grams of cardamom.

Since $100 is somewhat less than 20 times $5.59, we expect the answer to be somewhat less than 20 times 80 g. Since 20 times 80 g is 1,600 g, we expect an answer less than 1,600 g, so 1430 g is reasonable.

7. The cost to run the air conditioner for one hour is $3.60/24 = $0.15.
   The cost to run it 14 hours a day for one week is then $7 \cdot 14 \cdot \$0.15 = \underline{\$14.70}$.

8. If ten books weigh 27 lb, then one book weighs 2.7 lb. So the man can carry 90 lb / 2.7 lb = $33.\overline{3}$ books. Since he doesn't carry partial books, we need to round down and say he can carry <u>33 books</u>.

9. The area of lawn he needs to fertilize is 300 ft · 1,200 ft = 360,000 ft². So he needs to apply 360 lb of nitrogen (1 lb per 1,000 ft²). Although each bag of fertilizer weighs 10 lb, only 25%, or 2.5 lbs, of that is nitrogen. So he needs to buy 360 lb ÷ 2.5 lb/bag = 144 bags of fertilizer.

Puzzle corner:
Yes, it can. The proportion can be written in the form of $y = mx$ using the value 3/2 for m: $y = (3/2)x$. Notice that the proportion given is $x/y = 2/3$. This means that $y/x = 3/2$, from which we can easily solve for y to be $y = (3/2)x$.

Scaling Figures, p. 42

1. a. Example solution (proportion):

$$\frac{3}{5} = \frac{2.4 \text{ cm}}{x}$$

$$3x = 5 \cdot 2.4 \text{ cm}$$

$$x = \frac{12 \text{ cm}}{3} = 4 \text{ cm}$$

   b. Example solution: $x = 238 \text{ cm}/7 \cdot 3 = \underline{102 \text{ cm}}$

2. The scale ratio 3:4 means that the sides of the smaller triangle are 3/4 the length of the sides of the larger triangle. So we can just multiply each side of the larger triangle by ¾ to get the lengths of the sides of the smaller triangle:

   ¾ · (4.8 cm) = 3.6 cm,
   ¾ · (6.0 cm) = 4.5 cm, and
   ¾ · (3.6 cm) = 2.7 cm.

Scaling Figures, cont.

3. Answers will vary, but the corresponding angles of the two triangles must be equal, and the sides of the larger triangle must be 5/2 = 2.5 times longer than the sides of the smaller one. Please check the student's work.

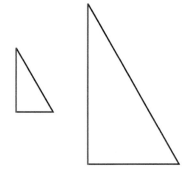

4. a.

	Length	Width	Aspect Ratio
Rectangle 1	1 cm	3 cm	1 : 3
Rectangle 2	1.5 cm	4.5 cm	1 : 3
Rectangle 3	2 cm	6 cm	1 : 3
Rectangle 4	2.5 cm	7.5 cm	1 : 3

   b. The aspect ratio for all four triangles is the same, 1:3.

5. a. x / 72 cm = 24 cm / 54 cm; x = 32 cm.
   There are three other ways to write a correct proportion and get the correct answer:
   72 cm / x = 54 cm / 24 cm ; x / 24 cm = 72 cm / 54 cm; and 24 cm / x = 54 cm / 72 cm.

   b. x / 43 m = 16 m / 20 m; x = 34.4 m.

   There are three other ways to write a correct proportion and get the correct answer:
   43 m / x = 20 m / 16 m ; x / 16 m = 43 m / 20 m; and 16 m / x = 20 m / 43 m.

6. a. Scale factor = *after* / *before* = 14 cm / 6 cm = 7/3 ≈ 2.33.
   b. Scale ratio = *after* : *before* = 14 cm : 6 cm = 7:3.

7. a. Scale ratio = *after* : *before* = 15 ft : 20 ft = 36 ft : 48 ft = 3 : 4
   b. Scale factor = *after* / *before* = 15 ft / 20 ft = 36 ft / 48 ft = ¾ = 0.75

8. If the area of the original square is 36 cm², then each side must be 6 cm long. Each side of the reduced square will be ¾ · 6 cm = 4.5 cm. So the area of the reduced square is (4.5 cm)² = 20.25 cm².

9. Please check the student's work. The size of the shape will vary according to how the page was printed. If the page was printed using a "scale to fit" or "print to fit" option, the actual measurements of the shape may not match what is given below. However, the scale ratio and the scale factor should be the same or very close, even if the page wasn't printed at 100%.

   The bottom sides of the two triangles measure 2.3 cm and 5.7 cm, so the scale ratio is 57:23. The scale factor is 57/23 ≈ 2.5.

10. Please check the student's work. The size of the shape will vary according to how the page was printed. If the page was printed using a "scale to fit" or "print to fit" option, the actual measurements of the L-shape may not match what is given below.

    The scale ratio 3:2 means the dimensions are multiplied by 3:2 = 1.5. The bottom width and the height of the L-shape both are 5.7 cm. These become 1.5 · 5.7 cm = 8.55 cm ≈ 8.6 cm.

    In inches, the bottom width and the height of the original L-shape are 2 ¼ in and become 1.5 · 2 ¼ in = 3 3/8 in.

    See the image on the right.

Scaling Figures, cont.

11. a. The sides are in the ratio: *after* : *before* = ¾" : 3" = 1:4.

   b. Let *x* be the second side of the similar rectangle. Since the *after* : *before* ratio is 1:4, the longer side of the second triangle is one-fourth the longer side of the first triangle: $x = 4\frac{1}{2}" \div 4 = 1\,1/8"$.
   You can also solve this using the proportion $x\,/\,4\frac{1}{2}" = \frac{3}{4}"\,/\,3"$; from which $x = 1\,1/8"$.
   These proportions work also: $4\frac{1}{2}"\,/\,x = 3\,/\,\frac{3}{4}"$ or $x\,/\,\frac{3}{4}" = 4\frac{1}{2}"\,/\,3"$ or $\frac{3}{4}"\,/\,x = 3"\,/\,4\frac{1}{2}"$.

   c. Area of the original rectangle: $4\frac{1}{2}$ in · 3 in = $13\frac{1}{2}$ in² = 13.5 in².
   Area of the similar rectangle: ¾ in · 9/8 in = 27/32 in² or 0.75 in · 1.125 in = 0.84375 in².

   d. The areas are in the ratio: *after* : *before* = 27/32 in² : $13\,1/2$ in² = 27/32 : 27/2 = (27/32) · (2/27) = 1/16 = 1:16.
   Or, using decimals, the ratio is *after* : *before* = 0.84375 in² : 13.5 in² = 0.0625 = 0.0625 : 1 = 1:16.
   So the ratio of the areas is the <u>square</u> of the ratio of the sides (square of the ratio 1:4).

Puzzle corner:
If the aspect ratio is 2:3, then the lengths of the sides are 2*x* and 3*x*. Thus the perimeter is
50 cm = 2*x* + 3*x* + 2*x* + 3*x* = 10*x*, so *x* is 5 cm. Therefore the sides are 10 cm and 15 cm long.

Shrinking the rectangle at a scale ratio of 2 : 5 is the same as changing *x* from 5 cm to 2 cm, so the sides of the
shrunken rectangle are 2 · 2 cm = 4 cm and 2 · 3 cm = 6 cm, and its area is 4 cm · 6 cm = 24 cm².

Floor Plans, p. 48

1. Please check the student's answers. They will vary according to what percentage of normal size the page was printed at.

   a. In the plan: 3/4 in by 1 1/2 in. In reality: 3 ft by 6 ft.

   b. In the plan: 1 in by 3/8 in. In reality: 4 ft by 1.5 ft

2. In the plan the room measures 3 in by 2 3/4 in. In reality, the room measures : 12 ft by 11 ft, so its area is 12 ft · 11ft = 132 sq ft.

3. A table that measures 3.5 ft by 2.5 ft in reality would measure 7/8 in. by 5/8 in. in the plan.

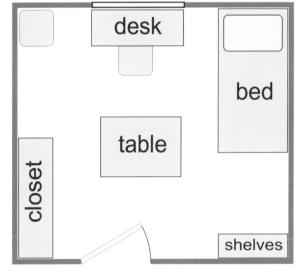

1 in : 4 ft

$$3\frac{1}{2}\text{ ft} \cdot \frac{1\text{ in}}{4\text{ ft}} = (7/2 \div 4)\text{ in} = 7/8\text{ in, and}$$

$$2\frac{1}{2}\text{ ft} \cdot \frac{1\text{ in}}{4\text{ ft}} = (5/2 \div 4)\text{ in} = 5/8\text{ in.}$$

4. a. To find the dimensions if it were drawn to a scale of 1 in : 6 ft, we first need to find the actual dimensions of the room and then to apply the new scaling.

   The actual dimensions are:
   $4\frac{1}{2}$ in · 3 ft / in = 13 1/2 ft by 4 in · 3 ft / in = 12 ft.

   Now apply the new scaling: 13 1/2 ft · 1 in / 6 ft = 27/12 in = 2 3/12 in = 2 ¼ in by 12 ft · 1 in / 6 ft = 2 in.
   So at 1 in : 6 ft the room would measure <u>2 ¼ in by 2 in</u>.

   An easier way to solve this is to recognize that a plan at 1 in : 6 ft is half the size of a plan at 1 in : 3 ft, so you can just divide the dimensions 4 ½ in and 4 in by two.

   b. To find the dimensions at a scale of 1 in : 4 ft, apply the new scaling factor to the actual dimensions of the room that you found in part (a):

   13 1/2 ft · 1 in / 4 ft = 27/2 ÷ 4 in = 27/8 in = 3 3/8 in, and 12 ft · 1 in / 4 ft = 3 in.

   So at 1 in : 4 ft the room would measure <u>3 3/8 in by 3 in</u>.

   Again, if you recognize that a plan at 1 in : 4 ft is ¾ the size of a plan at 1 in : 3 ft, you can just multiply the dimensions 4 ½ in and 4 in by 3/4.

5. Please check the student's work. The dimensions of the rectangle will vary according to what percentage of normal size the page was printed at.

    a. Printed at normal size, the plan should measure 5 cm by 7 cm. This means it is drawn to the scale
      5 cm : 2.5 m = 1 cm : 0.5 m = 1 cm : 50 cm = <u>1 : 50</u>.

    b. To calculate the dimensions for the plan, divide the true dimensions in centimeters by 50. On the plan, the windows should measure 80 cm/50 = 8/5 cm = 1.6 cm. The door should measure 100 cm/50 = 2 cm. The bed should measure 150 cm/50 = 3 cm by 200 cm/50 = 4 cm.

6. a. The dimensions on the house plan for the kitchen are 22.5 cm by 19 cm. That is, 4.5 m · 5 cm / 1 m = 22.5 cm, and 3.8 m · 5 cm / 1 m = 19 cm.

    b. In actual dimensions, the living room will be 5.2 m by 4.5 m. It's 26 cm · 1 m / 5 cm = 5.2 m by 22.5 cm · 1 m / 5 cm = 4.5 m.

7. Please check the student's answer. The room should measure 7.6 cm by 9.2 cm. The table in the middle should measure 2.4 cm by 2.4 cm

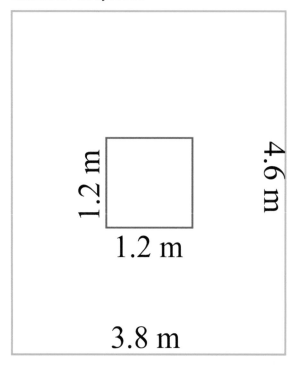

8. a. The plan should measure about 2 ½ in by 2 in, which corresponds to dimensions in reality of 20 ft by 16 ft. The area is then 20 ft · 16 ft = 320 sq ft.

    b. If drawn at the scale 3/8 in : 1 ft, each foot corresponds to 3/8 in. So the dimensions become 20 · 3/8 in = 60/8 in = 7 ½ in and 16 · 3/8 in = 48/8 in = 6 in. Since the rescaled width of the plan (7.5 in) exceeds the printable width of this page (7.1 in, printed at 100%), the rescaled plan isn't reproduced here.

Floor Plans, cont.

9. The plan measures 10 cm by 7.5 cm, which means that the true dimensions of the house are 10 m and 7.5 m. To redraw it at the scale 1 cm : 125 cm, we divide the true dimensions 10 m (= 1000 cm) and 7.5 m (= 750 cm) by 125:

1000 cm · 1 cm/125 cm = 8 cm, and 750 cm · 1 cm/125 cm = 6 cm.

The internal dimensions (such as the size of the bathroom) are scaled the same way. In fact, because the ratio of the two scales is 100 cm / 125 cm = 0.8, you can take any dimension in the original plan and multiply it by 0.8 to get the dimension in the new plan.

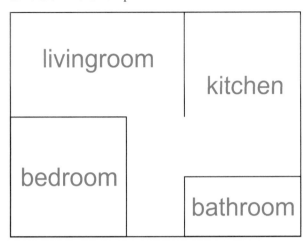

Maps, p. 52

1.

on map (cm)	in reality (cm)	in reality (m)	in reality (km)
1 cm	20,000 cm	200 m	0.2 km
3 cm	60,000 cm	600 m	0.6 km
5.2 cm	104,000 cm	1,040 m	1.04 km
0.8 cm	16,000 cm	160 m	0.16 km
17.1 cm	342,000 cm	3,420 m	3.42 km

2. a. 1 km; the scale is 1 cm = 1 km
   b. The actual length of the ski trail is 5.2 kilometers.

3. a. 1 cm is in reality 25,000 cm = 250 m = 0.25 km.
      So the scale becomes 1 cm = 250 m.

   b. See the table on the right.

on the map (cm)	in reality
2.0 cm	500 m
3.6 cm	900 m
6.4 cm	1.6 km
10.0 cm	2.5 km

4. Check the student's answers. The size of the map will vary according to the printer settings when it was printed. If the page was printed at "100% of normal size" (and not "scale to fit"), the answers should match the ones given below.

   a. The distance on the map is 3.5 cm. In reality, it is 3.5 · 180,000 cm = 630,000 cm = 6,300 m = 6.3 km.
   b. The distance on the map is 2.2 cm. In reality, it is 2.2 · 180,000 cm = 396,000 cm = 3,960 m ≈ 4.0 km.
   c. The distance on the map is 1.9 cm. In reality, it is 1.9 · 180,000 cm = 342,000 cm = 3,420 m ≈ 3.4 km.

5. a. At a scale of 1:500, 75 m is 7500 cm ÷ 500 = 15 cm on the map.
   b. At a scale of 1:1200, 75 m is 7500 cm ÷ 1200 = 6¼ cm on the map.

6. The distance 1.5 kilometers is 1,500 meters and 150,000 centimeters. Once we have converted the distance to centimeters, we can simply divide. For the map with a scale of 1:15,000, we calculate 150,000 cm ÷ 15,000 = 10.0 cm. For the map with a scale of 1:20,000, we calculate 150,000 cm ÷ 20,000 = 7.5 cm.

7. The nature hike is approximately <u>16 miles long</u>.
   In reality, the hike is 2.5 in · 400,000 = 1,000,000 in = 1,000,000 in · 1 ft/(12 in) · 1 mi/(5,280 ft) = 15.78$\overline{28}$ mi.

8. Please check the student's work.

9. Check the student's answers. If the lesson was printed, the size of the map will depend on how the printer scaled the printing. If it printed the page at 100% (and not "scale to fit"), then the answers should match the answers below. If it printed it at a different size, then the given scale on the map (1:50,000,000) is not correct. Even so, you can still check the student's answers, but the answers will neither match the ones below nor the distances in reality.

   a. The distance on the map from Tallahassee to Denver is 1 11/16 in. So in reality, it is 1 11/16 in · 50,000,000
      = 27/16 · 50,000,000 in = 84,375,000 in = 84,375,000 in · (1 ft)/(12 in) · (1 mi)/(5,280 ft) ≈ 1331.68 mi
      ≈ <u>1,300 miles</u>.

   b. The distance on the map from Sacramento to Austin is 1 7/8 in. In reality, it is 1 7/8 in · 50,000,000
      = 15/8 · 50,000,000 in = 93,750,000 in = 93,750,000 in · (1 ft)/(12 in) · (1 mi)/(5,280 ft) ≈ 1,479.64 mi
      ≈ <u>1,500 miles</u>.

   c. The distance on the map from Lincoln to Bismarck is 9/16 in. So in reality, it is 9/16 in · 50,000,000
      = 9/16 · 50,000,000 in = 28,125,000 in = 28,125,000 in · (1 ft)/(12 in) · (1 mi)/(5,280 ft) ≈ <u>400 miles</u>.

10.
   > First, I <u>divide</u> the distance 16.2 miles by the factor 500,000. I will get a very
   > small number, which will be in miles: 16.2 miles ÷ 500,000 = <u>0.0000324</u> miles.
   > Next I convert this to feet, and then to inches.
   > Converting miles to feet means to <u>multiply</u> by the ratio 5,280 ft/1 mi:
   >
   > $$\underline{0.0000324} \cdot \frac{5{,}280 \text{ ft}}{1 \text{ mi}} = \underline{0.171072}$$
   >
   > Then I convert the result from feet to inches by <u>multiplying</u> by the ratio 12 in/1 ft:
   >
   > $$\underline{0.171072} \cdot \frac{12 \text{ in.}}{1 \text{ ft}} = \underline{2.052864} \text{ in} \approx \underline{2.1} \text{ in.}$$

11. Since we will want to work on the map in inches, let's convert 45.62 miles to inches:
    First 45.62 mi · 5280 ft/mi = 240873.6 ft, and 240873.6 ft · 12 in/ft = 2,890,483.2 in.
    On a map with a scale of 1:250,000, 45.62 miles is 2,890,483.2 in ÷ 250,000 = 11.5619328 in ≈ 11 $^9/_{16}$ in.
    On a map with a scale of 1:300,000, 45.62 miles is 2,890,483.2 in ÷ 300,000 = 9.634944 in ≈ 9 $^5/_8$ in.

    (To convert decimal inches to a fraction:
    (1) Subtract the integer part: 11.5619328 in − 11 in = 0.5619328 in; 9.634944 in − 9 in = 0.634944 in.
    (2) Multiply by 16 (Fractional inches are measured in fourths, eighths, sixteenths, *etc*. It is unlikely that you will need to or even be able to measure much more accurately than a sixteenth of an inch):
        0.5619328 in · 16 = 8.9909248; 0.634944 in · 16 = 10.159104.
    (3) Round to the nearest sixteenth: 8.9909248 ≈ 9; 10.159104 ≈ 10.
    (4) If the number of sixteenths comes out even, then convert to eighths, fourths,
        or a half as is appropriate: $^{10}/_{16} = {}^5/_8$.)

12. In decimals 1 3/16″ by 2 1/8″ is 1.1875 in by 2.125 in. In reality the dimensions are
    1.1875 in · 15,000 = 17812.5 in by 2.125 in · 15,000 = 31,875 in. Converting to feet gives
    17812.5 in · 1 ft/12 in = 1484.375 ft by 31,875 in · 1 ft/12 in = 2656.25 ft.

    a. The (unrounded) area in square feet is 1,484.375 ft · 2,656.25 ft = 3,942,871.09375 ft^2.
       (Since the original data was accurate only to the nearest sixteenth of an inch, 3,900,000 ft^2
       would be a reasonable *rounded* answer.)

    b. The exact answer in (a) expressed in tenths of an acre would be:
       3,942,871.09375 ft^2 · (1 acre / 43,560 ft^2) ≈ 90.5 acres.

13. We can simply multiply the distance 5.0 inches by the ratio of the two maps' scales —
150,000/200,000 = 3/4 or 200,000/150,000 = 4/3. Since 1:150,000 is closer to reality
than 1:200,000 is, we expect the length of the hiking trail to be longer on the map at 1:150,000.
So we multiply the 5.0 inches by 4/3 to get 5.0 inches · 4/3 = 6 2/3 in ˜ 6.7 inches. (The original
length was given in tenths of an inch (5.0 in), so the answer should be to that same accuracy.)

   Another way to solve this is to first find the length of the trail in reality, and then find the length
of the trail on the map at 1:150,000. In reality, the trail is 5 in · 200,000 = 1,000,000 in
= 15.782828283 mi. On the other map, this distance will be 15.782828283 mi ÷ 150,000
= 0.000105219 mi = 6.666666667 in.

Puzzle Corner:
Let's calculate both dimensions, then compare them. The plot of land measures 1.65 km by 2.42 km, which
is 1,650 m by 2,420 m, which is 1,650,000 mm by 2,420,000 mm. Using the short sides gives a scale factor of
210 mm : 1,650,000 mm ≈ 1 : 7857. Using the long sides gives a scale factor of 297 mm : 2,420,000 mm ≈ 1 : 8148.
To make the map fit onto the paper, we have to choose the smaller scale, which is the bigger number, so the answer is
1 : 8148. (In reality, we would probably want to leave a margin unprinted around the edge of the paper, especially if we
were printing from an electronic printer that cannot print all the way to the edge, so we would be more likely to round
the answer to 1:8500 or even 1:9000 or 1:10,000.)

Significant Digits, p. 58

1. a. 3 digits   b. 3 digits   c. 4 digits   d. 1 digit   e. 2 digits   f. 5 digits
   g. 2 digits   h. 3 digits   i. 6 digits   j. 2 digits   k. 4 digits   l. 1 digit

2. a. Here we need to give the answer to 3 significant digits since both dimensions are given to 3 significant digits:
   24.5 m · 13.8 m = 338.1 m^2 ≈ 338 m^2

   b. Now both dimensions are given to 4 significant digits, so the answer is given to 4 significant digits also:
   24.56 m · 13.89 m = 341.1384 m^2 ≈ 341.1 m^2

3. a. Since 6.2 cm has two significant digits and the scale ratio has five, we give the final answer to two significant digits:
   6.2 cm · 50,000 = 310,000 cm = 3.1 km

   b. Since 12.5 cm has three significant digits and the scale ratio has six, we give the final answer to three significant digits:
   12.5 cm · 200,000 = 2,500,000 cm = 25.0 km

   c. Since 0.8 cm has one significant digit and the scale ratio has five, we give the final answer to one significant digit:
   0.8 cm · 15,000 = 12,000 cm = 0.12 km ≈ 0.1 km

4. The dimensions 5.0 cm and 3.5 cm are given to two significant digits. Let's calculate those dimensions in reality,
   and give them to two significant digits:

   5.0 cm · 8,000 = 40,000 cm = 400 m = 0.40 km

   3.5 cm · 8,000 = 28,000 cm = 280 m = 0.28 km

   The area needs also be given to two significant digits, since both numbers we multiply have two significant digits:

   A = 0.40 km · 0.28 km = 0.112 km^2 ≈ 0.11 km^2

5. a. 3.0 in · 10,000 = 30,000 in = 30,000 in· (1 ft)/(12 in) · (1 mi)/(5,280 ft) = 0.473$\overline{48}$ mi ≈ 0.47 mi

   b. 3.0 in · 10,000 = 30,000 in = 30,000 in · (1 yd)/(36 in) =833.$\overline{3}$ yd = ≈ 830 yd

6. Since 45.0 m and 21.2 have three significant digits and the scale ratio also has three, we will give the answers to
   three significant digits.

   45.0 m ÷ 500 = 0.09 m = 9.00 cm

   31.2 m ÷ 500 =0.0624 m = 6.24 cm

   The dimensions on the map are 9.00 cm by 6.24 cm.

1.

a. 41 km per hour	b. $\dfrac{3 \text{ g}}{800 \text{ ml}}$	c. $1 : 3$

2.

Miles	58	116	174	232	290	348	580	1,160
Hours	1	2	3	4	5	6	10	20

3. $20 \text{ g} : 1,200 \text{ g} = 1 : 60$

4. Susan jogs at a rate of $\dfrac{1\ 1/2 \text{ mi}}{1/3 \text{ h}} = \dfrac{3/2 \text{ mi}}{1/3 \text{ h}} = (3/2) \cdot (3/1) \text{ mi/h} = 9/2 \text{ mi/h} = 4\ 1/2 \text{ mi/h}.$

5.

a. $\dfrac{16}{17} = \dfrac{109}{T}$ $16T = 17 \cdot 109$ $16T = 1853$ $\dfrac{16T}{16} = \dfrac{1853}{16}$ $T \approx 115.81$	b. $\dfrac{1.5}{2.8} = \dfrac{M}{5}$ $2.8M = 1.5 \cdot 5$ $2.8M = 7.5$ $\dfrac{2.8M}{2.8} = \dfrac{7.5}{2.8}$ $M \approx 2.68$

6.

$$\frac{\$19}{12 \text{ kg}} = \frac{p}{5 \text{ kg}}$$

$$12 \text{ kg} \cdot p = \$19 \cdot 5 \text{ kg}$$

$$\frac{12 \text{ kg} \cdot p}{12 \text{ kg}} = \frac{\$19 \cdot 5 \text{ kg}}{12 \text{ kg}}$$

$$p = \$7.91\overline{6} \approx \underline{\$7.92}$$

7. Since $8 : 10 = 4 : 5 = 20 : 25$, Gary can expect to make __20__ baskets when he practices 25 shots.

8.

a. $\dfrac{2\ 1/2 \text{ pages}}{1\ 1/4 \text{ h}} = \dfrac{5/2 \text{ pages}}{5/4 \text{ h}} = (5/2) \cdot (4/5) \text{ pages/h} = 2 \text{ pages/hour}$ Alex solved problems at a rate of 2 pages per hour.
b. $\dfrac{2/3 \text{ room}}{3/4 \text{ h}} = (2/3) \cdot (4/3) \text{ room per hour} = 8/9 \text{ room/hour}$ Noah painted at a rate of 8/9 room in one hour.

Review, cont.

9. A car is traveling at a constant speed of 75 km per hour.

    a. $d = 75t$. See the plot below.

    b. The unit rate is 75 km/h.

    c. See the graph below.

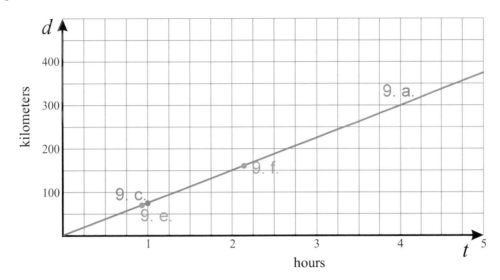

    d. The point (0, 0) mean the car has gone zero km in zero hours.

    e. Since 55 minutes is 55/60 = 11/12 hour, the car can travel $d = 75 \cdot (11/12) = 68.75$ km $\approx$ 69 km in 55 minutes. See the grid above for the point (11/12 h, 69 km).

    f. From the equation $75t = 160$ we get $t = 160/75 = 2.13333...$ hours $\approx$ 2 hours 8 minutes. See the grid above for the point (2.1 h, 160 km).

10. a. The quantities not in proportion. For example, 1 G for $10 gives us the unit rate of $10 per gigabyte, whereas paying $30 for 10G would give the unit rate of $3 per gigabyte. If they were in proportion, the unit rate would be the same no matter which two values you use to calculate it. You can also see it from the fact that whenever the bandwidth increases by 5G, the price increases sometimes $7, sometimes $6 so it does not always increase by the same amount.

    b. Does not apply.

11. It would cost $9,369 / 12 \cdot 5 \approx$ $3904 to drive the car for five months.

12. a. 8 m : 10 m = x : 6 m, from which x = <u>4.8 meters</u>. Or you can reason that the sides of the smaller triangle are 0.8 of the sides of the bigger, so the unknown side is 0.8 $\cdot$ 6 m = 4.8 m.

    b. 8 in : 5.6 in = 13 in : x, from which x = <u>9.1 inches</u>.

13. The true dimensions of the room are 2 in $\cdot$ 6 ft/1 in = <u>12 ft</u> and 2 ¾ in. $\cdot$ 6 ft/1 in = <u>16 ½ ft</u>.

Review, cont.

14. a. The unit rate is 6 mi/gal or 6 mpg (miles per gallon).

   b. M = 6*f*

   c. Answers may vary because the scaling on the axes may vary. Check the student's plot. For example:

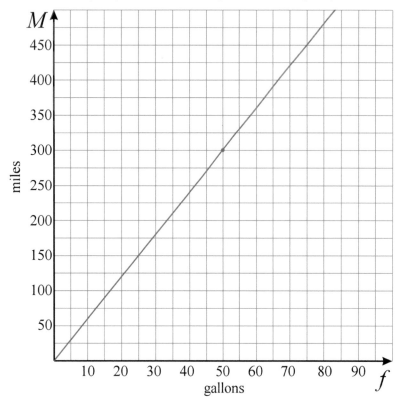